NATURE FAST AND NATURE SLOW

Nicholas P. Money is Professor of Biology and Western Program Director at Miami University in Oxford, Ohio. He is the author of popular science books on fungi and other microorganisms, including *The Amoeba in the Room: Lives of the Microbes* (2014), *Mushrooms: A Natural and Cultural History* (Reaktion, 2017) and *The Selfish Ape: Human Nature and Our Path to Extinction* (Reaktion, 2019).

Praise for *The Selfish Ape*

'I learned much from Nicholas Money's book. I love his vivid, prose-poetic imagery. Reading him is pure literary pleasure. He knows what to say and, more importantly, he knows how to say it.' – Professor Richard Dawkins FRS, author of *The Selfish Gene* and *Outgrowing God*

'A tour de force of life on Earth. Money eloquently describes the dynamics of life and the quite insignificant place of humans in the grand scheme of existence. Charting important biological discoveries, he describes life from all angles, including our molecular complexity and our genetic makeup . . . the book brings together many perspectives on human existence to create a beautiful but damning picture of humankind.' – *The Biologist*

Praise for *Mushrooms*

'A well written, authoritative and beautifully illustrated account of mushroom life and lore, leavened with humour. An ideal introduction to the most beautiful members of nature's least understood kingdom.' – Richard Fortey FRS, author of *Life: An Unauthorised Biography*

Praise for *The Amoeba in the Room*

'Nicholas Money is an expert guide . . . The world will not seem the same to anyone who reads his book.' – Helen Bynum, *Times Literary Supplement*

NATURE FAST AND NATURE SLOW

How Life Works,
from Fractions of a Second
to Billions of Years

NICHOLAS P. MONEY

REAKTION BOOKS

Published by

REAKTION BOOKS LTD
Unit 32, Waterside
44–48 Wharf Road
London N1 7UX, UK

www.reaktionbooks.co.uk

First published 2021
Copyright © Nicholas P. Money 2021

Printed and bound in India by Replika Press Pvt. Ltd

A catalogue record for this book is available from the British Library

ISBN 978 1 78914 404 8

CONTENTS

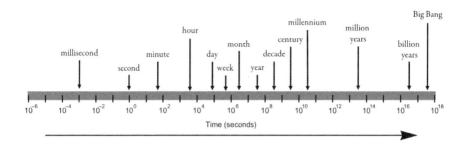

The arrow of time and the timescale of nature in seconds

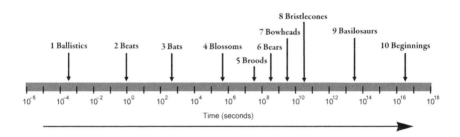

The timeline for the chapters in this book

PREFACE

Far away and far away
Flows the river of pure Day;
Cold and sweet the river runs
Through a thousand, thousand suns.

FREDEGOND SHOVE, *The River of Life* (1956)

Look in a mirror and remember your younger self. By middle age, the face of the youth hangs there, recognizable still, but camouflaged by blotches, creases and sags. The most expensive cosmetic procedures cannot conceal the truth from more than a fleeting glance. We get older without daily awareness of our passage. John Milton conveyed this with customary elegance in an early sonnet in which he cast time as 'the subtle thief of youth'.[1] That subtlety is everything. Time is missed so easily, clocked when our attention is on it, then flying onwards when life shakes us from its contemplation. Even when we are watchful, however, we are aware of just a sliver of it.

One second can be divided into 1,000 milliseconds or 1 million microseconds. Microseconds might as well be flying by in fairyland, but we are attentive to movements that occur in a few milliseconds.[2] From the Hawaiian shore, we marvel at the sight of a humpback whale breaching the water, and follow the steep curve of continuous motion and blunderous re-entry into the ocean. A video recording allows us to enjoy the titanic leap remotely on a mobile phone or television screen. Digital images captured at the standard cinematic speed of 24 frames per second are separated by 42 milliseconds (thousandths of one second). Watching the video, we perceive the movement of the whale throughout its breach, and we are unaware that individual images flash before us, like the pages of a flip-book, or cards upturned from a deck. The motion of the animal seems as smooth in the video as watching it live.

Some experiments suggest that the brain operates like a video camera, assembling movements as a series of discrete images to provide us with the illusion of continuous action. This feature of the underlying mechanism of vision is apparent from evidence showing that we can perceive the order in which distinct images are displayed when they are separated by as little as 3 milliseconds.[3] Hummingbirds beat their wings fifty times per second, which means that a full up-and-down sweep is completed in 20 milliseconds. This puts the motion of their wings at the edge of our awareness; the body of the hovering bird is sharp, while its wings fill blurred triangles. Looking at something faster, we notice the disappearance of a cat flea from one position and its seemingly instantaneous reappearance a pillow's-breadth away across a white bed sheet. In fact, a circus of hundreds of fleas could jump in quick succession, one after the other, in the time that elapses during the blink of an eye.

Human sensitivity to sound and to touch resides in roughly the same time frame as vision, with 2-millisecond audio bursts or needle-pricks making an impression on our consciousness and shorter stimuli escaping notice.[4] The overlap between these senses is not surprising, because the mechanisms that underlie all forms of awareness run on the same nervous hardware. Faster events can certainly kill us, but natural selection has not seen fit to engineer a sensation that would allow us to out-manoeuvre an oncoming bolt of lightning, cobra strike or stingray barb. Evolution spends its time working on the avoidance of more common causes of youthful death. This leaves us riveted to seconds and detached from an awful lot of life.

In the seconds that it takes to leave home in the morning, Eden – my garden in Ohio – is exploding with invisible movements, with a fusillade of spores bursting into the air from tiny fungi that have sprouted on rabbit pellets since sunrise, leafhoppers jumping from the car roof using gear wheels that spin faster than the pistons in the Porsche, and pond hydras firing barbs into their prey using pressurized harpoons. Does any of this matter to us? No more than the daily revolution of Earth, in the sense that we do not need to know that this Lilliputian circus is in full swing to get on with the day. Sunrise and sunset happened long before we understood that the Earth was spinning on its soft axle, rather than staying put and allowing the Sun to fling itself about us every day. Indeed, a survey conducted a few years ago revealed that as many as a quarter of Americans adhere to a classical view of the heavens.[5]

The uneducated may feel contented with the simplicity of their world view, or are unaware that there is more to know, but there is an awakening that comes from understanding that Earth is in motion and that life is lived on timescales that lie outside common comprehension. Mention of the cosmos is significant in this biology book, because a glimpse of the richness of nature lived in a single second is every bit as powerful as the sense of awe most of us feel when we look at the Milky Way on a clear night. As William Blake put it, 'If the doors of perception were cleansed every thing would appear to man as it is, infinite.'[6] The thrill of watching a high-speed video recording of a dragonfly rowing through the air on its stained-glass wings can be as stirring as seeing a solar eclipse.

These views of nature fast can be appreciated for their own beauty, without any need for us to ponder questions of wider importance. But there is also a practical side to studying the brilliance of fast movements, because nature fast is the foundation for nature slow. We cannot understand one without the other. Every second of the life of a body is populated with chemical reactions, sparking nerves and sliding muscle fibres. Scrutiny of these fast events is a vital part of figuring out how we work and how we might repair the damage resulting from accident or illness. Nature fast holds comparable significance for ecologists who study the biology of forest canopies or hydrothermal vents on the sea floor. Every organism is constrained by the speed of the chemical and physical reactions inside its cells. Working in the other direction, from slow to fast, the examination of slow nature helps us to understand why fast mechanisms work the way they do. Every exchange and transformation of molecules that keeps a cell running has to pass muster from generation to generation. Natural selection imposes itself as an inescapable filter that permits effective combinations of fast chemical reactions to flow through time and rejects anything that impairs survival. From microbe to whale, everybody lives on processes that play out over a range of timescales.

Bottom up and top down, we explore the entire timetable of the universe in this book, from slivers of a second to billions of years. After considering the fastest movements in nature, we turn to whole seconds. We mark time in seconds whenever we pay attention to the present, closing our eyes in moments of bliss to keep the feeling going a little longer, or, more often, waiting for seconds to pass, standing in line at the supermarket or sitting in a dentist's chair as the drill spins thousands of times per second, willing ourselves into the future. Minutes are more elusive, noted as soon as they have gone rather than as they pass. That is the way our perception of the environment is wired, and it comes from the biological need to make decisions from one second to the next. Nature seems to be attuned to this metre, although the precise length of a second is our invention, defined now by the frequency of an atomic clock. Our hearts beat every second – five times faster than the pulse of a hibernating marmot, and much slower than the twenty-beats-per-second flutter in the chest of a hummingbird. Elsewhere in the body, fluids ebb and flow every few seconds in the lymphatic and nervous systems, and bowel movements and orgasms obey similar pulse rates. Waves of much faster biochemical reactions control these and other second-by-second bodily fluxes, which illustrates the way in which we are swept into the future by the totality of natural processes that traverse several timescales.

Many of the behavioural programmes followed by animals are sustained for minutes and hours, and we have adopted hours as the main time intervals in our schedules. Daily or circadian rhythms are layered over the hours, and lengthier processes that depend on the accumulated effects of these activities orchestrate the life cycles of many animals, plants and microbes. The growth of plants and other photosynthetic organisms illustrates the link between the days and weeks of life. When sunlight bathes a forest of giant kelp, the flattened blades of these magnificent algae elongate by 0.5 m (1½ ft) per day, and on land, bamboo canes extend twice as fast, emitting an audible crackle. With the passage of months, seasonal changes in plant growth are evident in a cycle of greening that washes over the northern and southern hemispheres. Algal blooms cause widespread changes in ocean colour, too, particularly in coastal areas, where the seawater is enriched with nutrients that pour from the land. We mark months as they come and go, but sleep through one-third of the time and stay focused on the experiences of our conscious minutes and hours.

Processes that repeat on the basis of the elliptical track of the Earth around the Sun are described as circannual cycles. These include the migration of large herbivores, the winter hibernation of animals in cold climates, and the life cycles of annual plants and the pests that pursue them. Events that repeat over several years are not as common in nature, but the thirteen- and seventeen-year periodical cicadas of North America provide a spectacular demonstration of the way that development can be choreographed beyond a circannual cycle. Cicada larvae feed on tree roots, using a cryptic stopwatch to record the number of times sap flow ceases in winter or resumes in spring, and synchronize their metamorphosis into adults to this protracted pulse. This strategy of delayed emergence after a prime number of years helps the insects to outwit predators that cannot attune their reproductive cycles to the resurrection of the cicadas. When the cicadas emerge, the available predators are overwhelmed by their fantastic numbers and barely dent the cicada population before the adults have mated and deposited the next generation of eggs. The largest cicada broods are memorable events for those who experience the ear-splitting volume of their song.

Natural processes that appear to have some predictability or periodicity over decades emerge from underlying circadian and circannual cycles. Water bears, or tardigrades, are exceptionally resilient creatures that can survive heating and freezing, intense radiation, high pressure and tissue dehydration inside a vacuum. They have a global distribution and can be found, quite easily, by looking at strands of wet moss under a microscope. These simple animals enter repeated phases of suspended animation and can persist in a dried state for many years. They do not live for more than a year or two when they are active, but their ability to revive themselves after drying extends their lifespan to match that of much larger animals. Lungfish engage in similar behaviour by surviving drought encased in mucus cocoons, and can live for many decades. The life strategies of water bears and lungfish afford these species many opportunities to reproduce, despite their exposure to unpredictable and potentially lethal environmental conditions. The biology of these animals offers an interesting contrast with the human programme of continuous activity. This vivacity allows the oldest of us to live for 3 billion *seconds*, while our genes are the most recent versions of DNA instructions that have been touring Earth for 3 billion *years*.

We are mindful of the timescales of seconds through decades, but as this inquiry on time proceeds we begin to feel bypassed by the majesty of nature as she leaves us in the dust. Commercial whaling drove some populations of the bowhead whale to extinction, and came close to wiping out the species. Numbers have increased in the last half-century, and with an estimated 10,000 species of animals alive today, bowheads are no longer listed as endangered. It seems certain that the oldest of these highly intelligent animals recall the genocide that ended in the 1930s. A two-hundred-year-old animal caught by Inuit whalers in 2007 had the point of a nineteenth-century metal harpoon embedded in its tissue. Greenland sharks that swim slowly in the frigid waters of the Arctic Ocean may live twice as long as bowheads. These sharks have been aged by radiocarbon dating the eye lenses of animals caught by marine biologists. The accuracy of this method is proven by the observation of a pulse in radioactivity in animals born after the atmospheric tests of nuclear weapons in the 1950s.

Many trees outlive humans, and none does so by longer than the famous bristlecone pine of California, Nevada and Utah, whose lifespan reaches 5,000 years. There are much older tree species, including the 80,000-year-old quaking aspens of Utah, but these are clones within a colony of plants that interconnect through their roots. Single stems may not last for very long at all. Colonies of Mediterranean seagrass are even older, raising the possibility that there is no inherent limit to the lifespan of some plants, as long as their environment remains stable. This is complicated by the way that colonial species are continually refreshed by new growth as older parts die, which means that we must rethink the nature of the individual and how we should measure age. Pond hydras, whose barbs were mentioned earlier, caused a sensation among natural historians in the eighteenth century when it was shown that they could regenerate from snippets of their tentacled bodies and form natural buds that would go on living indefinitely. Some of this sense of wonder can be recaptured in our time when we contemplate the history and development of the ancient colonies, or mycelia, of honey mushrooms.

Evolutionary changes can be monitored over short periods, particularly in studies that show the adaptation of bacteria and other microorganisms to environmental challenges. But when we think about big changes in biology, including the transformation of marine species into land animals, and vice versa, we must look at the deep history of groups of organisms over tens of millions of years. Turtles left dry land and freshwater habitats to splash into the sea at least four separate times over the last 260 million years. Seagrasses made the same marine transition from several groups of ancestral plants 100 million years ago. The 8-million-year rise of whales from hoofed mammals related to the hippopotamus was a more recent and very speedy metamorphosis.

Turning our attention from millions to billions of years, we intersect with the origin of life, the creation of our solar system and the history of the universe. The passage of time on this universal scale challenges our comprehension as much as the movements that are over in a microsecond. It seems plausible that life arose as soon as the physical conditions on Earth allowed the precursors of biomolecules to remain stable – although this may have happened repeatedly, with several intervening planetary sterilizations, before the ancestors of today's biology took hold. Once the first cell, or protocell, began replicating, the growth of a founding population of microbes would have occurred in an explosive fashion on an uninhabited planet. And life would have begun changing from the very beginning according to the all-pervading law of natural selection. This view of an accelerated commencement of life encourages the idea that various forms of biology may be commonplace in the universe. The importance of fast events within the immensity of time brings the book full circle, uniting the time frames of the ten chapters and situating our lives in a fresh and rather glorious perspective.

Each chapter in this book explores a particular slice of time, beginning with nature fast, spanning millionths to tenths of one second, and progressing to nature slowest, occupying billions of years. Using the mathematical notation of powers of ten, Chapter One is concerned with the swiftest movements of organisms that are accomplished in 10^{-6} to 10^{-1} seconds, and Chapter Ten covers the beginnings of life 10^{17} seconds ago. The timescale of the book gauged in seconds covers 24 orders of magnitude.[7] If we were measuring distance rather than time, this span of numbers would stretch from the length of a bacterium to the distance to the stars in our galactic neighbourhood. As we look at the way biology works over the lengthier timescales, the narrative relies more on the past tense than when recounting the mechanisms in the present. This reflects the way that we interact with the rest of nature – always in the now – and the imaginative challenge of examining our place within the larger and older structure of life. The future tense is also essential for considering the affairs of the biosphere once *Homo sapiens* makes its exit and after the eventual vaporization of Earth.

There is a deliberate void at the beginning of this journey across time frames. The movements of molecules within cells take a few thousandths of a second, and the faster chemical reactions and some unconscious neurological processes are accomplished in millionths of a second. These mechanisms are incorporated into the first chapters. But some chemistry happens much more swiftly, including the transfer of energy across chlorophyll molecules when they are warmed by sunbeams. The methods used by some enzymes to accelerate reactions inside cells fall into the same category that we call quantum events. These take place in femtoseconds, which are quadrillionths of a second, or 10^{-15} seconds in our power-of-ten notation. Quantum processes underlie everything, in a sense, because they govern the behaviour of atoms, but they have been ignored by most biologists until very recently. We thought we could examine nature without bothering about the playfulness of atoms and subatomic particles, but this has begun to change with the recognition that quantum events may help to explain how life works.[8] *Nature Fast and Nature Slow* is sufficiently expansive already without dwelling on this largely theoretical field of quantum biology, but it is good to acknowledge this mysterious foundation of time. We begin with things that happen 1 billion times more slowly, namely the fastest natural movements that can be seen when we slow the action using high-speed cameras.

Portuguese man-of-war, *Physalia physalis*, a siphonophore rather than a true jellyfish, composed of integrated colonies of individual animals called zooids.

ONE

BALLISTICS

Fractions of Seconds (10^{-6}–10^{-1} Seconds)

Evolution invented life's fastest movement on the tentacles that trail from jellyfish bells. The result was the nematocyst, an explosive single-celled harpoon, used to snag brine shrimp and other sea creatures. Human swimmers are stung by these weapons on sea wasps and sea nettles, lion's mane jellyfish and Portuguese men-of-war.[1] Nematocysts borne on the Irukandji box jellyfish deliver a venom of such nastiness that these pea-sized animals threaten tourism on the coast of Queensland.

The nematocyst is a plump, poisonous urn, with a barbed dart poised beneath a hatch and a connecting tube coiled in its base. When this microscopic gadget is triggered, the hatch flips open, the dart is ejected and the tube uncoils. Poison flows through the tube and the swimmer who brushed against a box jelly and received a thousand stings collapses on the beach. Dart movement is fast, done in a microsecond (10^{-6} seconds); tube discharge takes as long as a millisecond (10^{-3} seconds). The dart reaches a top speed of 70 km/h (44 mph) and slams into skin, or the shell of a shrimp, with the ferocity of a rifle bullet. It travels only a short distance from the open hatch, but reaches an acceleration of millions of g as it zips across the gap to its target.[2]

Biology could not dream of making a nematocyst for the first billion years of life. Back then, evolution was limited to customizing the structure of sea-dwelling bacteria, which could never accommodate a weapon of this sophistication. The difficulty for the bacterium lies in its formulation as a blob of fluid surrounded by a single membrane. Shooting a needle from a cell made like this would be suicide – like popping a balloon. To construct a nematocyst, a separate internal locker is needed to hold a tube that can be emptied without rupturing the whole cell. This accessory arrived with the invention of eukaryotes, whose cells import and export goods – solid food in, waste materials out, and all manner of secretions secreted – within compartments sealed with their own membranes. Nematocyst discharge is a very violent form of secretion.

Jellyfish stingers are powered by the release of hydrostatic pressure, which seems to be the best way to make a tiny projectile move very, very fast. The pressure, which is estimated to be as high as 150 atmospheres, is generated by the accumulation of a propellant that dissolves in the fluid inside the nematocyst. This causes water to diffuse into the cell, by the process of osmosis, which has the effect of pressurizing the stinger and priming it for discharge. The chemical compound that pressurizes the stinger is called polyglutamic acid, or PGA. Another place that we find PGA is in the sticky, fragrant goo of fermented bean pastes, including kinema from eastern India and the better-known natto from Japan. (I adore Japanese cuisine, but find this substance nauseating.) It seems that the vital gene in the jellyfish DNA that allows them to make PGA in their stingers came from marine relatives of the bacteria that we use to make bean pastes. We refer to instances like this, of DNA transfer between different species, as horizontal gene transmission.

Pond polyps, or hydroids, related to jellyfish deploy the same weapons. These simple animals stick to aquatic plants and wave their tentacles in the water, which accounts for the name hydroid, referring to the many-headed serpent of Greek mythology. They are like miniature upside-down jellyfish. Nematocysts are also used by corals and sea anemones, which, together with jellyfish and hydroids, form a large grouping, or phylum, of animals called cnidarians.

Nature bristles with poisonous stingers, from the noxious hairs of nettles to the stings of wasps and bees. Stingrays use spines with serrated edges and venom-filled grooves for defence. They arch their tails to expose their spines and jab trespassers with a swift upward lunge. The Australian conservationist and television presenter Steve Irwin was killed by a huge stingray in 2006, joining Odysseus in the short list of individuals slayed by these cartilaginous fish. (Odysseus was killed with a spear tipped with a stingray spine, fulfilling the prophecy in the *Odyssey* that his death would come from the sea.[3]) Matching the unconscious inventions of nature, humans settled on stinger designs to hunt whales, from handheld spears in the Neolithic to harpoons fitted with grenades in the nineteenth century. Returning to microscopic lances, a profusion of nematocysts evolved in things unrelated to jellyfish, including single-celled protists (which we used to call protozoa and algae), parasitic slime moulds that poke holes in plant cells, waterborne fungi and relatives of the pest responsible for the Irish potato famine. By engaging in an arms race against their prey species, these microbes have fashioned some amazing contraptions that make the jellyfish missile seem rather plebeian. One weapon, of mind-boggling complexity, is produced by an ocean-going microbe called a dinoflagellate. Not content with firing a single dart, this multi-barrelled nematocyst shoots a dozen of them from a magazine. It is a miniature Gatling gun, and an individual dinoflagellate cell is like a submarine equipped with a battery of these astonishing armaments.[4]

Evolving on branches of the tree of life that lacked nematocysts, sea slugs, flatworms and comb jellies elected to steal them from jellyfish and sea anemones and deploy the needles on the surface of their bodies to discourage predators.[5] A spectacular sea slug called the blue dragon achieves this heist by eating the Portuguese man-of-war, coating the hijacked stingers in slime to prevent them from firing, and moving them to little skin sacs. Anything that starts chewing on the sea slug is stung. Decorator crabs attach sea anemones to their carapaces and acquire protection from their stingers as the anemones benefit from carriage to new feeding areas on their mobile allies. Nematocysts have also been harvested by medical researchers who use them to sting their patients. In clinical tests, they have isolated the stinging cells from sea anemones, stabilized them by drying to produce a powder, and mixed this into a gel that is applied to the skin of human volunteers. Each treatment contains millions of nematocysts. When a solution of a water-soluble drug is added to the gel, the nematocysts fill with the medicine as they rehydrate, and eject it into the skin. Tests show that these preparations are more effective at administering drugs used to prevent motion sickness and other conditions than transdermal patches.[6]

Nematocysts were discovered in 1702 by the Dutch scientist Anton van Leeuwenhoek, who saw them through one of his handheld microscopes. He described their globular shape on pond hydras, but had no inkling of their purpose. The Swiss naturalist Abraham Trembley went a little further in the 1740s, drawing hairs extending from tentacle surfaces without recognizing that they were shot from the globules. Research on nematocysts was stymied by the poor quality of microscopes, and their real function was not understood for another century, until corrected lenses were introduced that produced crisper images. In between, nematocysts were described as 'testicles', because the tubes dangling from discharged cells were interpreted as sperm tails, and, closer to the truth, 'anglers', because they were seen as retractable fishing lines. The mist cleared as research continued, and three hundred years after the first glimpse of a nematocyst, a high-speed video camera running at more than 1 million frames per second allowed us to watch them firing.[7]

High-speed video cameras are also needed to catch fungi in the act of catapulting, spurting and flipping their spores into the air. These are the only mechanisms that compete with nematocyst discharge for the title of Nature's Fastest Movement. Spores are the equivalent of seeds for a fungus. As a young researcher, I became fascinated by the work of a handful of investigators on the way that spores become airborne. A mushroom cap releases 30,000 of these microscopic particles every second, casting fairy dust into the wind. Spores are dotted over every patch of a mushroom gill, so that under the microscope the magnified gill looks like a field of watermelons. And then one of the spores vanishes and reappears instantaneously a short distance away in the air, followed by another, and another, until the gill is empty. Each spore appears to show the quantum quality of being in two places at the same time – on and off the gill – but the trickery comes from its speed. The launch is achieved with a microsecond catapult, making the lift-off too sudden to be grasped in the milliseconds perceived by the brain.

My predecessors tried to capture the flying spores using cine cameras fitted with motorized film canisters. The motors roared as they ripped thousands of frames of photographic film past a microscope lens in a second. When the film was developed, the researchers would be teased by the image of a single spore, sitting still in thousands of frames, then exasperated when it disappeared without a trace, leaving a colossal length of film with blank frames. The spore jumped invisibly in the instant between two frames. Time was on my side a few decades later, when advances in video technology brought microsecond imaging to the lab. When the cameras used to study nematocysts were turned to the fungi, the views were stunning: the mushroom spores employed a 'surface-tension catapult', powered by the motion of water droplets, to fling themselves from their gills.[8]

Fungi that do not produce mushrooms show comparable wizardry in their movements, and there is a parade of inconspicuous species that contribute to a festival of invisible gymnastics. A fungus that grows on horse dung uses pressurized liquid in its translucent stalk to propel a spore-filled capsule more than 2 m (6½ ft). Another species, called the artillery fungus, can fly more than 6 m (20 ft). The fruit bodies of this species grow in wet patches beneath trees and shrubs, and resemble mustard seeds before they germinate. In preparation for the launch, the little spheres of the artillery fungus crack open to form a cup that holds a glistening spore capsule. This cup is an elastic membrane that flings the capsule into the air by flipping outwards. The capsule travels at 10 m/s (33 ft/s), or 36 km/h (22 mph), so we might spot its flight if we could maintain focus on the speck as it flies silently above the grass. The effectiveness of this mechanism is apparent from the freckles on my wife's car, which are shot from a flower bed next to her parking space at work. The fungus is phototropic, meaning that it responds to sunlight, so it points its fruit bodies towards the reflection of the morning sun in the silvery sheen of the car. Complaints about this microbe and tips for removing its sticky blobs occupy thousands of recent web pages, suggesting that the fungus – which adores warm, wet weather – may be a beneficiary of climate change.

The fastest fungus is a microscopic species that is adapted for growing on vegetation after it has been scorched by fire. This microbial athlete is known as red bread mould, which is a lamentable name for a fungus that is orange and is a rarity on stale loaves. It forms thin 'torpedo' tubes that become pressurized like nematocysts. These are forced open when a plug is expelled from an aperture at the tip, and release four spores, one after the other, at a top speed of 115 km/h (72 mph). The emergence of the spores is not quite as fast as the jellyfish stinger, but all four of them are in flight within a few microseconds and they travel a distance of 1 cm in half a millisecond. The reason fast launch mechanisms like this have evolved lies in the physical resistance of air. Small things experience air as a viscous fluid, and spores would not fly more than a couple of times their own length if they were shot at slower speeds. By exceeding 100 km/h, the fastest fungus is able to travel five hundred times its own length. The flight of the bread mould succeeds in getting it into the air, where it can be spread much further by wind gusts.

The movements nature packs into such thin slivers of time are powered by the release of stored energy. Nematocyst detonation is a perfect example of this kind of mechanism. Before the stinger detonates, the fluid contents of its cell become pressurized with its propellant. When the hatch flips aside, the pressure is released and pushes the dart forwards. The nematocyst has a high level of potential energy before it is triggered, and this is converted into the kinetic energy of the moving dart when the hatch opens.[9] The torpedo tubes of the red bread mould work in this way, too. They are pressurized by a biochemical process that extends over many minutes, and this slow loading of chemical energy into the tubes contrasts with the rapid release of pressure and expulsion of the spores. The amazing launch speeds of these devices are achieved by this mechanism of power amplification, which is comparable to stretching a crossbow string before the release of the arrow.

The principles at work here are more obvious when we look at the behaviour of animals that amplify the power of their muscle contractions to execute a movement. The fastest muscle movements include the 98 km/h (61 mph) sprint of the prong-horn antelope of North America. I have startled these beautiful mammals on the shortgrass prairie of Colorado, and watched with amazement as they race across the flat landscape, kicking up clouds of dust. Their agility is thought to have evolved as they evaded American cheetahs, which, being extinct, no longer trouble them. This fascinating idea is based on the sprinting speed of the living, but endangered, African cheetah, which clocks a top speed of 100 km/h in three seconds with a monumental burst of millisecond muscle contractions. Another impressive volley of muscle contractions allows a woodpecker to peck tree bark twenty times per second, or once every forty milliseconds, which matches the fastest human drumbeat and tap dance.[10] Muscles cannot move limbs any faster, but their power can be amplified using natural springs and catapults.

To boost their acceleration, fleas, crickets, grasshoppers, froghoppers and other insect acrobats use springs made from a compound called resilin, a highly elastic protein that behaves like natural rubber. In a relaxed state, the chemical strands that form this substance arrange themselves as a tangled ball. Preparing to jump, fleas lock their hind legs in position, and squeeze a tiny pad of resilin at the base of each leg by contracting muscles that pull on their external skeleton. The resilin strands straighten when they are compressed, creating an ordered structure that is poised to spring back into the disordered ball. The instant the legs are unlocked, the resilin pads expand, transmitting their stored energy into the jump, and the insect bounds into the air in a thousandth of a second. Froghoppers – also known as 'spittle bugs' because their nymphs protect themselves in a froth of 'cuckoo spit' – launch themselves even faster. Another group of bugs, the planthoppers, have a jump mechanism that uses toothed gears to ensure a smooth take-off by forcing their hind legs to extend in synchrony.[11] The continuous reworking of the external skeleton of insects to permit these jumps, and the associated restringing of muscles and rewiring of nerves, is a marvel of evolutionary adaptation.

Other power-amplified movements occur in spiders and shrimp. Trap-jaw spiders, which live in New Zealand and South America, have large jaws, or chelicerae, tipped with curved fangs. They hold these wide open and then snap them together in as little as one-tenth of a millisecond. Mantis shrimp use spring-loaded claws to crack shellfish, and have shattered aquarium glass in captivity. The claws strike at over 100 km/h, more than twice as fast as the record karate chop – which is more impressive when we consider that the shrimp is moving its claws against the resistance of water. The movement is so forceful that it creates a stream of gas bubbles in the water akin to the action of a speedboat propeller.[12] The bubbles collapse very quickly as the gas is squeezed by the pressure of the surrounding water, creating heat, a flash of light and a shockwave that amplifies the wallop delivered by the claw strike.

Ferns use bubble explosions to cast their spores from microscopic onagers – Roman catapults – clustered on the underside of their fronds.[13] Onagers were wooden siege engines designed to fling rocks at enemy forts. The arm carrying the rock in a sling was forced down against the resistance of twisted ropes, which provided the necessary power amplification. The fern version of the onager has a backbone of elasticated cells, which curls backwards as it dries out, forming a springy cup that cradles a dozen or more spores. Tension in the backbone increases as it is bent by the continuing evaporation of water, until the explosive expansion of a gas bubble in each cell allows it to snap forwards in ten microseconds, scattering the spores into the air. Pollen is flung from white mulberry flowers at a similar speed by the recoil of the elastic filaments that support the anthers holding the grains.

Aquatic plants called bladderworts reverse the common purpose of fast movements, using tension within their tissues for suction rather than expulsion. Their pretty flowers, which look like little snapdragons, grow on aerial stalks, and peppercorn-sized traps sprout on the stems that dangle in the water. The traps expel water, creating a negative pressure as the walls of the structure are pulled inwards. Trapdoors bristling with trigger hairs close the traps. When a hapless water flea brushes against these hairs, the door flips open, the victim is sucked inside and the trapdoor slams shut again in half a millisecond. Unable to move, the prey is asphyxiated as it runs out of oxygen, and is digested by the carnivorous plant. The relatively sedate closure of the Venus flytrap, which involves the release of tension in its opposing leaflets, takes a tenth of a second.[14]

Some lightning-fast movements are audible. In *The Sea Around Us* (1951), the American conservationist Rachel Carson described the 'crackling, sizzling sound, like dry twigs burning or fat frying', associated with colonies of pistol shrimp that snap their claw pincers together.[15] These crustaceans have one outsized claw and one more modest appendage, so that they look like boxers wearing single gloves. The sound is produced by clacking the pincers of the big claw, like castanets, propelling a jet of water that leaves a trail of bubbles. The pincers are hinged and can be cocked after muscle contraction, storing energy that produces the accelerated movement when they are released, and resulting in a shock wave that the shrimp use to stun fish. The cacophony produced by these animals is one of the loudest sounds in the ocean.

The launch of the artillery fungus can be heard without amplification, and the cannon of the fastest fungi are audible as a fizzing sound when thousands of them clustered in a cup fungus discharge at the same time. All the rapid movements described in this chapter create sound waves. We cannot hear nematocysts, but the immense number of jellyfish and their relations, and all the microbes with stingers, make it possible that one of them explodes every microsecond of every day, causing the continuous vibration of water. If extraterrestrial astronomers have trained their radio telescopes on Earth and tuned them between megahertz and kilohertz frequencies – 1 million to 1,000 cycles per second – our galactic neighbours might have heard an accelerating string of faint pings after the rise of the eukaryote cell, until the jellyfish filled the oceans and Earth resounded with the discharge of darts and the unslithering of their poisonous tubes. The jellyfish recording would be the first movement of the symphony of nature fast, followed by fungal water guns, snapping ferns and leaping insects, a billion-year performance of pops, spits, whirs and boings.

Scientists in every field are pushed to assess the value of their work for the rest of humanity. The customary justification for studying fast movements lies in the potential for translating discoveries about the machinery of natural mechanisms into commercial products. The use of jellyfish stingers for administering drugs is one example of a potential application. Other investigators have suggested that a synthetic form of flea resilin could be used to produce artificial knee joints, and that latches and springs inspired by insects could be incorporated into miniaturized drones. Bioengineers are committed to these technological goals, but many biologists are content with unveiling the beauty of natural designs. The problem, of course, is that the rest of society is paying for this research.

After I gave a presentation to a group of mushroom enthusiasts, one member of the audience asked me to share more about my research into fungal spores. He told me that he worked as a sanitary engineer and had been interested in mushrooms since childhood. I described the torpedoes of the red bread mould, and as my explanation proceeded it became apparent that he was beginning to view me as a French peasant might have regarded Marie Antoinette. On frosty mornings when he had peered into septic tanks, I had watched the black pips of a fungus streak like meteors across the circular field of the microscope lens. Neither of us was impressed by the argument that his blind investment in my work, through his taxes, was justified by the observation that fungi are instrumental in decomposing waste in septic tanks.

A better defence rested on the sheer beauty of some of the fast movements captured with high-speed video, whose surprising gracefulness, when displayed on a laptop, seemed to inspire my acquaintance as much as me. In other words, the possibility for enlightenment grows when we recognize that some of this research pulls us into an arena that dissolves the distinction between science and art.

Humans have always known that flea jumps are very fast, from the simple observation of the disappearance of the parasites as we try to catch them beneath a fingertip. Our familiarity with these inflammatory pests is reflected in the glorious Baroque paintings that show young women in various states of undress catching fleas.[16] Contemporary readers were enthralled by the pull-out illustration of a flea magnified to the size of a cat in Robert Hooke's masterpiece *Micrographia*, published in 1665. Members of the Royal Society were fascinated by the anatomical details revealed by Hooke, but wider interest came from the emotional impact of this novel representation of a tiny animal. So much of nature fast remains fantastically arcane. Nematocysts are felt but not seen, most often as a tingling sensation when we brush our fingertips against a sea anemone in a rock pool. The speed of the jabs is inconsequential.

The fast movements of animals, fungi and plants happen as snippets of activity within the lives of organisms that extend over terrifically longer periods of time. Jellyfish outrigged with lightning-fast harpoons can live for years, and some may edge towards immortality, as we shall see later. This feature of life, as something that is expressed over multiple timescales, will impress itself throughout this book. The muscles and nerves that control an insect jump grow and connect over days of assembly inside the pupal soup, and none of this takes place at all unless the animal survives the larval phase before its miraculous metamorphosis. At the tighter end of the timescale, the necessary chemical components of muscles and nerve are crafted through millisecond chemical reactions that operate as steps in the vast web of metabolic cycles swirling throughout the life of the insect. If we could watch the chemistry – see every reaction occurring within a second in the tissues of an insect – we would be overwhelmed by the turmoil of molecular exchanges. But somewhere in the depths of this tempest arise the signals for the flea or the leafhopper to cock its legs, squeeze its elastic skeleton, open the latch and rocket into the air. And this is life: a magical carnival of chemistry and physics, born from chaos and experienced in movements fast and slow.

TWO

BEATS

Seconds (10⁰ Seconds)

What happens now is not seen, heard or felt for a tenth of a second. When we catch up, the immediate event is already part of the history of the universe. Have you ever looked at a clock and thought that the second hand had stopped ... and then watched it tick on? This is the experience of chronostasis, evoked by a mismatch between visual stimulus and perception – between seeing something and knowing that you saw it. The clock had not stopped, of course, but it took your brain a moment to process its movement. Take a deep breath. Your brain sent nerve impulses to your chest muscles a fraction of a second before you felt your nostrils flare. There is another short delay between nostril expansion and consciousness of the air flowing into your lungs. More generally, feelings of pleasure and pain build in split seconds before we are mindful of either sensation. We are unaware of the present until it becomes the very recent past. Consciousness is fed to us from the unconscious.

The runner sizes up the woodland stream and decides to jump before she is aware of thinking, 'I can do it!' Most of the time we are oblivious to these interruptions, which are programmed into the nervous system. Such time lags are essential for survival because they ensure that the brain has sufficient time to gather information arriving from different circuits, increasing the probability that we will make a suitable response. The process was uncovered in the 1970s by Benjamin Libet, a physiologist at the University of California, who recorded unconscious brain activity urging decisions before the conscious decision to act.[1] These observations, which have been confirmed by subsequent research, suggest that free will is a myth – an idea that disturbs people who like to think that humans are more than the sum of their animal parts. Broader critiques of free will rest on common experience and a wealth of psychiatric studies that reveal the overarching power of the subconscious over behaviour. We know that the seemingly conscious way the axe-murderer begins swinging the blade is predetermined by an unfathomable, but not infinite, list of existing factors. The fact that he is not wholly in control of his actions does not make his victims feel any better, or affect the judicial penalty for his behaviour. Enslavement to the subconscious is not a legal defence.

Having acknowledged these limitations in human comprehension, it certainly feels as if we live in the present, skipping from one immediate moment to the next. Even if we accept that this is an illusion, it does not make much difference to how we conduct our waking lives. A fly lands on your nose; you brush it away *immediately*, or swiftly enough. When we pay attention to the passage of time we count seconds, or watch them tick by on a watch face.[2] Fractions of a second pass too swiftly for us to tally them, and minutes flow too slowly. Seconds mean everything to us, but they do not exist by themselves. We invented them in the seventeenth century, when pendulum clocks provided the necessary precision to divide day-lengths into 86,400 intervals.[3] The choice to divide hours and minutes into sixtieths is rooted in Sumerian and Babylonian mathematics, which used sexagesimal numbers, or base 60, in their counting systems.[4] Since 1967 the second has been defined by a particular large number of vibrations of non-radioactive caesium atoms, as their outermost electron moves between high and low energy states billions of times per second. Seconds are measured by atomic clocks tuned to the frequency of microwave energy emitted by the caesium atoms as they undergo these 'hyperfine transitions'.

The human heart beats every second, sometimes faster, sometimes slower, but not much faster or slower. When we hold our breath, this internal metronome forces itself into our consciousness, and the pulse may be the reason that we fixed on seconds as a useful measure of time in the first place. Heart rates are related to body size. Comparison of the smallest and largest mammals makes this point: the heart of the Etruscan shrew has the weight of a grain of rice and beats up to 25 times per second (25 Hz), whereas the 200-kilogram (440 lb) pump inside the chest of the blue whale contracts only eight to ten times per minute (0.1–0.2 Hz). The *lub-dub* of the whale heart can be heard for 3 km (almost 2 mi.) underwater, with the tricuspid and mitral valves closing to keep the pressurized blood in the ventricles on the *lub*, and the aortic and pulmonary valves closing when the blood is on its way through the gigantic aorta to the lungs on the *dub*. The anatomy of the whale heart is surprising only in its size. The layout of the shrew heart is similar, but it stops fluttering after a year or two, compared with the eighty years of service delivered by the pump in the whale. In terms of the lifetime number of heartbeats, however, the shrew and whale organs show similar endurance. This encourages the idea that all mammals are allotted a maximum number of heartbeats and live fast for a year, or slow for a century.

The oldest individuals of most species experience an average of 1 billion heartbeats, although we can squeeze 3 billion from ours.[5] 'The flame that burns twice as bright burns half as long,' as the Chinese philosopher Lao Tzu is said to have remarked.

The beating of the heart originates in the contraction of muscle cells concentrated in the right atrium (upper chamber) in a structure called the sinoatrial node. These cells pulse up to one hundred times a minute, 'firing' in response to the movement of charged atoms, or ions, of sodium, potassium and calcium through their membranes. They do this automatically, even when they are separated from other cells in a plastic tissue-culture dish. The movement of the charged ions causes changes in the electrical potential, or voltage, between the inside and outside of the cell, in much the same way that electrical impulses are stimulated in nerves. Each muscle cell is connected to its neighbours through special junctions, so that the electrical stimulus associated with contraction spreads across the whole heart. In this manner, the

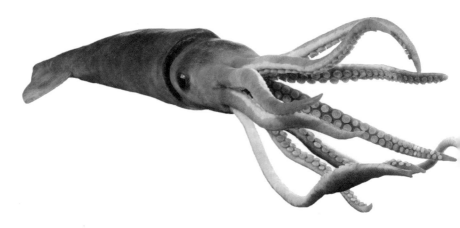

Giant squid, *Architeuthis* species, which, like other kinds of squid as well as octopuses, possesses three hearts.

sinoatrial node acts as a natural pacemaker, and its effect on heart rhythm is transmitted to cells in a secondary node and spreads to the walls of the ventricles (lower chambers) through a system of fibres. The heart is hooked up to the nervous system, which adds another layer of control to the activity of the pump, increasing or decreasing its pace according to the need, respectively, to hit the panic button or have a snooze.

Simple hearts are found in earthworms and are little more than a series of swellings along blood vessels that arch around the front end of the gut. Molluscs have more complex circulatory systems, ranging from a two-chambered heart in garden snails to a trio of separate hearts in cephalopods. The beating heart of a snail can be seen through its shell with bright illumination, and pulses every second or two. Two of the hearts in squid and octopuses pump blood to the gills. The third organ is flushed with blood returning from the gills, which is enriched with oxygen and distributed to the rest of the body with each pulse. Despite this bizarre vasculature, squid have the same resting heart rate as humans.

As a young scientist, I spent a brief spell experimenting on squid, which required me to hold the animals in my hand and feel them writhe before decapitating the poor creatures with a pair of razor-sharp surgical scissors. This was necessary to isolate the fat nerve cell called the giant axon, which is very useful for studying the transmission of nerve impulses. There was no value for a student in this foray towards vivisection. The workings of the 'action potential' can be grasped by anyone with a modicum of intelligence and access to a physiology textbook: sodium out → depolarization, potassium in → repolarization, followed by over-shoot and recovery. Did my victims feel the quickening pulse in my fingers before squirting ink in a desperate bid for life? The fields of neurology and cardiology have rested for centuries on animal torture.[6]

Insects have an open circulation in which their pale blood, or haemolymph, is pumped through a vessel that runs along the back of the animal, beneath its exoskeleton, and is expelled into the body cavity. A contractile heart at the abdominal end of this vessel pushes blood towards the head, which floods the rest of the body cavity and is drawn back into the heart through paired intake valves as it relaxes between beats. Some insect hearts can reverse flow to squeeze blood directly into the abdomen. Smaller contractile vessels, or auxiliary hearts, pump blood into the wings and antennae. This circulatory system supplies tissues with nutrients, but the role of gas exchange is subcontracted to the system of tracheal tubes that connect every part of the animal to the outside air via perforations in the exoskeleton. These tubes are effective for insects of different body sizes, from microscopic fairy flies to the mouse-sized larvae of goliath beetles.

Spiders have an open circulation, like insects, but their haemolymph has the added task of supplying their hollow limbs with hydraulic power. Compression of the haemolymph in the thorax raises its pressure and extends the legs; relaxing this pressure allows the leg muscles to pull the limbs closer to the body. This explains why the limbs of a spider curl inwards in death.[7] The relationship between heart rate and body size found in mammals applies to spiders, too.[8] Small jumping spiders have heart rates above one hundred beats per minute; tarantula hearts beat only ten times per minute. Over this size range, jumping spiders and other active hunters tend to have faster hearts than species, such as the highly venomous brown recluse, that make crude webs and wait, with an air of insouciance, indulging their arachnid dreams until insects become entangled and require attention.

As circulatory systems have been modified in different groups of animals, nature has been compelled to work with a limited selection of body parts to solve problems in fluid mechanics. Whether or not a set of blood vessels carries food alone, or oxygen alongside the dissolved nutrients, pumps are essential, and they must feed into blood vessels that diminish in size until they are fine enough to service individual cells. As evolution has unfolded, hearts have gained chambers, and shrunk and ballooned according to the exigencies of body size and behaviour. The invention of the giraffe presented some notable challenges, but the carotid arteries lengthened and blood pressure increased to prevent dizzy spells in a brain raised 2 m (6½ ft) above the heart. Yet, with all the anatomical changes wrought by natural selection, the fundamental structure of vascular tissues has never varied all that much. From these ground rules of nature, hearts that beat every second or so evolved in worms and spread to slugs and spiders, fish and beyond.

In line with this conservation of vascular engineering, we find that the genetic underpinnings of heart development and the mechanisms that control the heartbeat are universal. Fruit-fly larvae that hatch with mutations in a particularly important gene do not develop a heart, a fact that led investigators to name the gene 'tinman'. Tin Man, who wanted a heart, was the most sympathetic character in *The Wizard of Oz*, and expressed a Buddhist reverence for all of biology.[9] A version of the same gene, identified with the less poetic name *NKX2.5*, is found in humans. Mutations in its code result in malformations during foetal development, including holes between heart chambers and abnormal positioning of the aorta. Thankfully, many of these congenital defects can be fixed by paediatric surgery.

Cardiac pacemakers are widespread among animals, with islands of muscle cells thumping away even when the heart is unhooked from the nervous system. In a few instances, this device is a curse. Insects and spiders that are stung with nerve toxins by parasitic wasps lose control of their muscles, which depend on stimulation by nerve impulses, but are kept alive, in a paralysed state, because their hearts continue to beat. This gruesome fate provisions the wasp larvae, which hatch from eggs laid on the helpless host, with plenty of fresh food. In a similar vein, hearts cut from pithed fish and reptiles edify biology students by continuing to beat for hours, and hearts torn from sacrificial victims throbbed as they were offered by Aztec priests to their god, Huitzilopochtli. All nervous connections are severed in hearts removed for transplantation, but the intact pacemakers in the donor organs sustain the pulse on their own in their second homes. Transplanted hearts are not as responsive to exercise or excitement, but beat on, boats against the current, at one hundred or more beats per minute.[10]

Newborn babies have fast pulses and breathing rates: 120–160 beats per minute and a breath every second, or second and a half. These signs, along with muscle tone, irritability expressed in spirited crying and comprehensive pinkness, predict a vigorous, albeit post-traumatic, future. Years later, lying grey rather than suffused with pinkness, the chest stops heaving, and the opinion 'He's gone' rests on the absence of a pulse. William Wordsworth hit on the nuts and bolts of life in his poem 'She Was a Phantom of Delight' (1805):

> And now I see with eye serene
> The very pulse of the machine;
> A Being breathing thoughtful breath,
> A Traveller betwixt life and death.[11]

We pulse – all nature pulses – betwixt life and death. Wordsworth's metre is iambic tetrameter, as he used in 'I **wan**/dered **lone**/ly **as**/ a **cloud**.' Each iamb is a metrical foot of paired syllables, and the bold type highlights the stressed syllables of each iamb. Iambic pentameter adds an iamb, as we can hear in the first line of Shakespeare's Sonnet 12: 'When **I** / do **count** / the **clock** / that **tells** / the **time**.' Iambic hexameter, referred to as the alexandrine, was used by Edmund Spenser in *The Faerie Queen* (1590) as the wrap-up line for each stanza of his Elizabethan epic, following eight lines of iambic pentameter.

Dactylic hexameter is the form used in classical Greek and Latin poetry – Homer and Virgil – and was adopted by some of the practitioners of modern rap music. Dactyls are built from one long syllable followed by two short (unstressed) syllables. Alfred, Lord Tennyson used pairs of them in 'The Charge of the Light Brigade' (1854), which begins: 'Half a league, half a league, / Half a league onward.' Stringing six of these together per line does not work particularly well in English poetry. It is more effective in German verse, and researchers from Germany and Austria (no bias there, then) have suggested that reciting poems in dactylic hexameter may be valuable in the treatment of speech impediments and difficulties in understanding speech caused by brain injury. Evidence for this comes from research showing that the recitation of translated verses from the *Odyssey* or *Iliad* resulted in the harmonization of the pulse rates and breathing patterns of the study participants.[12] The resulting relaxed breathing frequency of six breaths per minute and heart rate of sixty beats per minute may have some value in the field of 'anthroposophical therapeutic speech'. The heart beats as the *Iliad* begins, 'Sing now *lub-dub*, goddess *lub-dub*, the wrath *lub-dub* of Achilles *lub-dub* the scion *lub-dub* of Peleus *lub-dub*.'[13]

The cadence of circulating blood is connected to the slower flow of the colourless liquids of the lymphatic and nervous systems. Lymph is formed from the fluid derived from blood plasma that bathes our tissues. It is filtered as it passes through the lymph nodes, and is charged with white blood cells that protect us from infection. Up to 5 litres (10½ pints) of lymph flow through the lymph capillaries and vessels every day before draining into the bloodstream. There is no steady rhythm to the motion of lymph, but, rather, irregular contractions of larger vessels that last for a few seconds and squeeze the fluid along.

Cerebrospinal fluid has a pulse, too. This clear, salty liquid, derived (like lymph) from blood plasma, washes through the central canal of the spinal cord and the brain ventricles. It also occupies the space between two of the three membranes, or meninges, that protect the outer surface of the brain and spinal cord. Half a litre (1 pint) of this thin broth is produced every day, but the quantity that is circulating through the central nervous system is limited to the volume of a large glass of wine by continuous reabsorption into the bloodstream.[14] Different rhythms have been recorded in this fluid, with slow tidal waves that play out over minutes, as well as faster movements occurring at up to eight sloshes per minute. These seem to be linked to heart contractions, but the motion of cerebrospinal fluid is separated from the bloodstream by a tissue barrier and is quite remote from the regularity of the heartbeat.

The smooth muscles of the digestive system pulse, from slow contractions of the stomach every twenty seconds, to faster waves of intestinal squeezing that arrive every five seconds. Like the heart, the digestive system is connected to the brain via the autonomic nervous system, the circuitry that controls the unconscious activities of life. Half a billion nerve cells form sleeves around the digestive system, all the way from beak to arsehole, as one might put it. These nerves form an interconnected web, or second brain, that controls the blood supply and release of digestive enzymes, and choreographs the all-important waves of contraction – one-way peristalses at the front end and two-way mixing movements called segmentations closer to the rear end. The first and second brain are tethered via the vagus nerve, but the digestive system will continue to pulsate, just like the heart, if this is severed. Peristaltic waves are stimulated by pacemaker cells in the gut wall, and they contract with, or without, a conversation with the head. This does not mean that the second brain is very clever, however; mine is an irritable little chap with few skills in problem-solving.

From the persistent peristalses of the bowel, we come to the brevity of the orgasm, both male and female versions. Anal and vaginal probes were used in classic experiments at the University of Minnesota in the late 1970s to record the pelvic contractions in volunteers who pleasured themselves to orgasm.[15] Averaged over many volunteers and merging different categories of orgasm, women experienced one pelvic contraction every two seconds for 36 seconds, and the anal probe measured a single contraction every second and a half for 26 seconds in men. The accompanying ejection of semen is a high-velocity event, with a launch speed of 45 km/h (28 mph), slightly faster than the discharge of the artillery fungus featured in Chapter One, but falling well short of its 6-metre (20 ft) trajectory. Sticking with the reproductive system, slow contractions of the uterus occur during the menstrual cycle, increasing from one or two per minute as follicles in the ovaries mature, to twice that rate at ovulation. Menstrual cramps also follow an irregular rhythm as the uterus sheds its lining.

The contraction of muscle cells in the heart, lymph vessels, spinal fluid, digestive tract and reproductive organs is driven by molecular motors that attach to the surface of internal protein filaments. As these motors grab and release, like fishermen hauling on nets, the filaments slide along one another and pull against the membrane at opposite ends of the cell. Shortening and fattening of the muscle cell as the filaments slide together, followed by relaxation as they slide apart, is controlled by waves of biochemical reactions that criss-cross the cell. Fresh proteins are made, from start to finish, in ten seconds, but much of this chemistry belongs to the domain of milliseconds or even faster, as the arrow of time flies on. All the while, the cells coordinate their contractions, causing the muscles to pulsate, not too fast, not too slow, second by second, pushing the fluids around, urging food through the gut and forcing blood around the body. When we concentrate, we feel the heart beat, more as a single *bub* than as the *lub-dub* heard with a stethoscope, each one pushing us into the future. A second quotation from Wordsworth is fitting here, this one from *The Prelude*, in a loose sort of iambic pentameter:

> With life and nature, purifying thus
> The elements of feeling and of thought,
> And sanctifying by such discipline,

Both pain and fear, until we recognize
A grandeur in the beatings of the heart.[16]

Feeling my heart beat, there seems no question that the past, present and future are real chunks of time. A Cooper's hawk landed on my garden fence a few minutes ago. It was watching my chickens pecking on the lawn, and weighing its options. The white bantam, Rosie, is a perfectly sized meal for the wild bird. I opened the window and the raptor flew away. There was a past, during which Rosie might have been carried off; then there is a present, when the predator flies away; and now Rosie has a future. Some philosophers and physicists disagree with this innocent reading of events.

Philosophical presentism is the belief that the present is the only time that exists. This seems reasonable, if a little restrictive. The future does not exist, yet, and past events are not happening any longer, so, yes, only the present is real, however fleeting. Presentism contrasts with eternalism, or the block universe theory, which suggests that there is no objective flow to time and proposes that the past, present and future coexist. This conflicts with the experience of living: I am writing this paragraph at home today, and it is snowing; I spent yesterday travelling; I am not sure what I will be doing next week, whether the snow will have melted or even, unfortunately, if I will be alive. Eternalists like to think about multiverses, which afford the dimensions for a cosmic circus of all possible worlds and events. Who knows, perhaps I am a banana talk-show host in an alternate universe, interviewing my hero, John Milton, who happens to be a stick of rhubarb. Finally, the *growing* block universe theory accepts that the past and the present exist, but sees the future as a pliable thing whose action will occur soon and then become part of the historical archive. This last idea seems the most sensible one because it corresponds to our familiar description of time.

Even if we cannot embrace the physical theories that recommend eternalism, it is clear that *now* is a relative term and that my present is not the same as the present on a planet orbiting another star. Astronomers training their staggeringly powerful telescopes on my garden from one of the exoplanets orbiting Kepler-446 would observe a family of Native Americans planting corn in the seventeenth century. Considerably closer to home, anyone living on Proxima Centauri b, our closest exoplanet, who wishes, or wished, to send friendly greetings to me today, would need to have transmitted their message via radio waves four Earth-years ago. But whether we sit by a fireplace in a living room here or anywhere else in the universe, our hearts will beat and the 2 billion seconds of an average lifespan will pass to the tick-tock of the caesium atoms in the clock on the mantelpiece. For a second is always a second.

THREE

BATS
Minutes and Hours (10²–10³ Seconds)

H e stood beside his car for a few seconds, feeling the chill of the pre-dawn air and watching the white cloud of his exhalation disperse into the darkness, then drove off. The rest of the day passed without the conscious attention to a single moment, until just before he fell asleep, when he felt his life beat on, second by second, from the rhythm of his pulse. Years later, in the last weeks of his life, breathing was difficult, making what had been reflexive unpleasantly present, so that each second became a painful intrusion into his sleepless minutes and hours. His daughter visited often in the afternoons and read aloud some of his favourite poems. He laughed when she put on a silly voice. It was then that he became indifferent to the strain of breathing and time seemed to pass more swiftly until she left.

Whether we embrace or endure the experience of being alive, we count seconds, but the chain of events, memorable or not, is stretched over minutes and hours. Virgil commented on this slippage of time as we are distracted by the immediate, in the *Georgics*:

> *Sed fugit interea, fugit inreparabile tempus,*
> *singula dum capti circumvectamur amore.*[1]

Little brown bat, *Myotis lucifugus*, a North American
mouse-eared microbat.

> But meanwhile it flees, time flees irretrievably,
> While we wander around, prisoners of our love of detail.

Behavioural episodes last for an age compared with the movements
we considered in the first chapter, but they are compiled from the
accumulation and interplay of these unconscious chemical and
mechanical processes. From late spring through to autumn, little
brown bats flutter in the sky above my garden. They come from
their roost under the oak-beamed roof of a covered bridge that
crosses a creek at the edge of the neighbourhood. The size of mice,
but bearing just one-third of their weight, these agile mammals
flap their membranous wings eight times per second and sweep
through the air, repeating elongated figures of eight every minute,
with frequent diversions in pursuit of midges, mosquitoes, moths
and beetles in the warm Ohio sky. These insects are located from
the rebound of sonar pulses emitted from the bat larynx. Bats
decode these echoes to create three-dimensional images of their
auditory space, including the positions of the insects, and increase

the number of calls to keep track of their prey as they close in for a kill. This hunting behaviour is called hawking, although the top speed of the bat is less than half the speed of the Cooper's hawk that eyes my chickens.

A handheld ultrasonic detector allows us to listen to the bats by bringing the frequency of their echolocations into our range of hearing. The detector converts the energetic calls into a clatter like a tin drum, which intensifies as a bat flies towards the detector and stops dead as it strikes its prey, crunches the insect with its sharp teeth and resumes its sweeping circuit between the trees. Munching on a fat moth must be like eating a cheese-filled pastry, with the crunch of the chitinous exoskeleton giving way to the creaminess of the tissues inside. Or is it?

'What Is It Like to Be a Bat?' is the title of a controversial essay published by the American philosopher Thomas Nagel in 1974.[2] Nagel argued that it is not possible for us to comprehend the experience of life for a bat, because the experience of being one – or any other animal – cannot be reduced to phenomena that can be measured scientifically. His ideas are taught today in philosophy classes by professors who cling to the idea that there is an inherent barrier between the mechanism of the mind and scientific laws. They elide the obvious absence of an alternative supernatural explanation for consciousness, and do not explain why the experience of life, by bats or humans, requires anything other than the electricity and narcotics provided by the nervous and endocrine systems. Nagel and his modern acolytes seem stuck with the phantom of the mind–body dualism recommended by their seventeenth-century progenitor, René Descartes.

While we cannot replace our awareness with the intelligence and deep emotional life of a little brown bat, we are in a very good position to understand something of that experience.[3] Nagel was particularly strident in his exposition of echolocation as something beyond human comprehension:

> bat sonar, though clearly a form of perception, is not simi- lar in its operation to any sense that we possess, and there is no reason to suppose that it is subjectively like anything we can experience or imagine.[4]

This is an argument without a sliver of merit. For starters, evolution has built bat sonar 'from the "standard" mammalian auditory system', which means that we have an alternative version of the same hardware.[5] Although we cannot picture the sky through our ears, I can get an inkling of this enhanced sense by closing my eyes and listening to bat calls with my detector. Blind devotees of echolocation can go much further in their appreciation of the chiropteran experience, visualizing their surroundings from the return of tongue-clicks and negotiating city streets with evident aplomb.[6] Echolocation 'is not magic', as one prominent neurobiologist has noted, which is one more reason why philosophers must abandon the hand-waving that affords them the space to treat the mind as something that exists beyond the material world.[7]

Echolocation by bats is a high-speed vocalization that relies on some of the fastest muscle contractions in nature. Rattlesnake rattles and toadfish grunts are the only sounds produced by muscle movements with similar speeds.[8] Little brown bats emit up to two hundred calls per second; each of these notes is held for a couple of milliseconds and has a maximum frequency of 80 kHz. We cannot hear sound waves at this frequency, or anything above

20 kHz, and many of my generation cannot get anywhere close to this with the inner-ear damage caused by live rock concerts in the 1970s – where the sound seemed perfectly balanced *outside* the arena after it was filtered through layers of masonry.

Behavioural routines for bats and other animals run for minutes and hours, and this is the metre that dominates the study of animal behaviour, or ethology. To continue this exploration of what it is like to belong to another species, we can begin to examine the emotional lives of little brown bats. They flit to and fro, minute after minute, sometimes for hours without stopping for a hang from the rafters. It seems likely that they enjoy their nightly hunting missions, feeling conscious from second to second during flight and receiving a pleasurable burst of serotonin with a successful insect strike. What about the experience of the prey? Antlions are among the rarer delicacies enjoyed by the bats above my garden. These insects spend most of their lives as larvae called doodlebugs, which excavate pits in sandy soil. They are caught by bats when they emerge as adults from their pupae and take flight as a type of lacewing. Antlions would seem to be helpless against their much faster sonar-equipped predators, but they can evade capture if they hear the approaching bat a second or two before they are hawked from the air. Startled by a sonar burst, antlions freeze their wings in a horizontal position, flex the abdomen to touch the underside of the thorax, and pitch into a nosedive.[9] The antlion behaves like a fighter pilot who lowers the elevators on the tailplane to send the aircraft into a dive. How antlions hear bat sonar is a mystery, for now. Related insects detect bat ultrasound via a pair of fluid-filled swellings on their wings that are fitted with tiny 'eardrums', and similar structures may be buried somewhere among the delicate tracery of wing veins of the adult antlion.

The doodlebug phase of the antlion can last for two or three years, and most of it is spent waiting, patiently, for opportunities for extreme violence. A worker ant going about her business makes a random error by walking too close to the edge of a pit. She slips and tries to crawl upwards, and then the flicking begins. Sand grains shower her – the size of house bricks relative to the ant – causing an avalanche, and she falls back into a pair of curved jaws, which snap together, piercing her crust, and she is dragged below and darkness closes her eyes.[10] The antlion is triumphant. Its first meal in a month. Second by second the larva fights the ant, both animals pushed into overdrive by the rush of a hormone called octopamine (the insect equivalent of noradrenaline), and the antlion is rewarded with minutes of gustatory pleasure. Afterwards, the humdrum life of the antlion continues. The larva waits, and waits, and waits for the slightest movement of sand that signals a visitor. The brief struggle must be fantastically rewarding for the nervous system of the insect. Why wait, otherwise?

Tiger moths are less cowardly than the nosediving antlions, choosing to keep flying and flexing the little drumskins on their thorax to produce their own ultrasonic clicks to jam the bat sonar.[11] Confused by the interference of these rapid clicks, the bats miss their targets and the moths keep flying, laughing at their hapless predators. We can keep going down this wormhole, of sorts, from exploring what it is like to be bats, antlions, ants and moths, to other species, elaborating this description of life to the whole tree of it, and discovering, as the naturalist John Muir said, that 'when we try to pick out anything by itself, we find it hitched to everything else in the universe.'[12] But one more organism hitched to bats deserves mention before we move on. This is the fungal pathogen that causes white-nose syndrome and has brought the little brown bat from 'least concern' in 2008 to 'endangered' in 2018 by the International Union for Conservation of Nature. The spread of this pest and the grievous decline in insect numbers are driving the little brown bat to extinction. In this instance, there is a blessing to feel like a bat, because these sensitive creatures have no conception of the coming apocalypse. They live in the present – or as close to it as the inbuilt delay in their nervous systems allows – anticipate sunset and leaving the roost, and think nothing about tomorrow, as far as we know. Seconds to live by, minutes and hours for behaviour, and little concern for the future.

Toothed whales are the marine masters of sonar. Sperm whales, which have the largest brains, employ a highly sophisticated version of the standard auditory technology to hunt squid. They make clicking sounds by forcing air through one of their nasal passages and closing a pair of valves, called monkey lips, on the forehead. When they smack their lips together, the sound waves are projected backwards through the oil tank called the spermaceti organ, where they hit an air sac and are bounced forwards through a second oil-filled chamber called the junk, and out into the water. Sound waves returning from a squid, which may be more than half a kilometre (⅓ mi.) away, are transmitted to their internal ears via a fat pad at the back of the jaw and are decoded by the 8-kilogram (18 lb) brain. They use a buzz phase in their pursuit of calamari that is redolent of the clicking of bats in their final approach to a moth. Unlike bat clicks, however, the whale vocalizations fall into the audible range for humans.

At a maximum intensity of 200 decibels, a whale could destroy a diver's eardrums, but we do not swim anywhere close to the depths where they click for squid; scuba divers rarely exceed a depth of 50 m (165 ft), whereas routine hunting by sperm whales involves a thirty- to forty-minute plunge to a depth of 400 m (1,300 ft). They reach depths of 2 km (1¼ mi.) in dives of exceptional duration.

Sperm whales use the same sound system for communicating with other whales, but we have little idea what they are saying, other than, 'I'm over here' and 'Where are you?' Beyond our ignorance of their language, there are many ways in which the experience of being a whale seems at least as alien to us as the life of a bat. The continuous passage of whales through liquid and the deep, three-dimensional nature of their marine habitat contrast with the two-dimensional focus of our excursions. As strange as we perceive the experience of whales, they may wonder also at our freakishness. Herman Melville pondered this in *Moby-Dick* (1851):

But far beneath this wondrous world upon the surface, another and still stranger world met our eyes as we gazed over the side. For, suspended in those watery vaults, floated the forms of the nursing mothers of the whales, and those that by their enormous girth seemed shortly to become mothers. The lake, as I have hinted, was to a considerable depth exceedingly transparent; and as human infants while suckling will calmly and fixedly gaze away from the breast, as if leading two different lives at the time; and while yet drawing mortal nourishment, be still spiritually feasting upon some unearthly reminiscence;

– even so did the young of these whales seem looking up towards us, but not at us, as if we were but a bit of Gulfweed in their new-born sight. Floating on their sides, the mothers also seemed quietly eyeing us.[13]

Differences between the way that organisms compile visual images present other challenges for answering the question 'What is it like to be a _____?' Insects are capable of processing more images per second than humans. While we sit enthralled by the talk-show host on television, a dragonfly alighting on the arm of our couch would yawn at the slow flipping of still pictures, opting to clean her antennae rather than wait for the on-screen divinity to stop smirking at his own brilliance. Dragonflies enjoy a 'flicker fusion frequency' of more than 200 Hz, meaning that they can perceive more than two hundred distinct images in a second. Flicker rates vary across vertebrate animals, too, with smaller species gathering more information than larger ones. This has led to the impression that time may seem to pass more slowly for smaller animals, which enjoy more action in a few minutes than we pack into an hour.[14] This is one of those features of animal behaviour that is difficult for us to comprehend, but not impossible. The Scottish contemporary artist Douglas Gordon created *24 Hour Psycho* in 1993 by slowing Alfred Hitchcock's movie of 1960 from 24 frames per second to 2 frames per second, extending the viewing experience from 109 minutes to an entire day.[15] The famous shower scene, which lasts for two and a half minutes in the original, is extended for half an hour in Gordon's version. We view his film as a dragonfly would experience the original, and share their fascination, no doubt, with the beauty of Janet Leigh. (Although, slowed to two frames per second, Leigh's higgledy-piggledy teeth become very noticeable.)

Videos of slime moulds do not benefit from a slo-mo edit. There is nothing to slow. These organisms evoke the passage of minutes and hours better than anything, because they do very little at shorter or over longer intervals of time. Their name is the embodiment of sluggishness, and slime they do. Against this anthropocentric slander, and in their considerable favour, we should note that they have been sliming around for hundreds of millions of years, are very colourful and beautiful, and will be here long after humanity is fossilized. The common name refers to a single stage in a life cycle of these organisms, which is formed when microscopic amoebae fuse to produce a conspicuous glistening ooze that migrates over rotting wood consuming bacteria, fungal spores and other scraps of organic matter. This is called a plasmodium, and the largest ones can envelop a tree stump. These giant amoebae are vascularized with channels that spread like blood vessels, from larger tubes, or 'veins', at the rear of the plasmodium, to tiny 'capillaries' that reach the pulsating edge of the colony.

If you look at an active plasmodium in the woods, you may notice that it is shimmering. This delicate agitation of its surface is caused by the constant motion of fluid within its veins. Cellular fluid, or cytoplasm, containing fat globules and sparkling granules streams through the veins at a speed of 1 millimetre per second, and reverses direction every minute or minute and a half. A low-power microscope reveals the extraordinary animation of this organism, with the propulsion of fluid across the field of view in one direction, then a pause, followed by a reversal of flow. Energy is distributed across the whole plasmodium through the veins, from the older parts of the slime mould to the leading edge that fans out at a speed of a few millimetres an hour. The flow of cytoplasm in the veins is driven by the movement of protein filaments that form a skeleton inside the slime mould. This bears comparison to muscle contraction at the molecular level. Nothing like this happens in the leaf veins of plants. Water and dissolved sugars move through leaves without any minute-by-minute reversals in direction.

Slime moulds belong to a large group of organisms that includes species of *Amoeba* that live as single-celled microbes and glide over decomposing plant fragments in ponds as miniaturized versions of the pizza-sized plasmodia. All these amoeba-like organisms are more closely related to animals and fungi than they are to plants, which does not seem surprising when we contrast their mobility with the stationary lives of plants. As plasmodia explore their environment in search of food, they flow around barriers and make choices about the optimal way to channel themselves through their veins.[16] If one side of the plasmodium finds itself rippling over a site that has high levels of nutrients, it makes sense for the whole organism to move in that direction, rather than continue to spread where nothing has been found. Once the food is exhausted in one place, the slime mould can resume its quest by redirecting its fluid flow until it finds a fresh food supply somewhere else.

To increase the probability of locating food, the plasmodium does not waste time running back over areas that it has already covered. It 'knows' how to do this by avoiding surfaces that are marked with its slime trail, which serves as a kind of simple mapping method. Plasmodia can also be trained to traverse an obstacle laced with chemical repellent if they are rewarded with food afterwards. This is an example of habituation. They can even communicate this information to a neighbouring plasmodium when they make contact, so that the second slime mould learns to overcome its innate aversion to the irritant in favour of receiving the food reward. Unlike the slime trail, which signals 'AVOID ME' for the rest of the life of the slime mould, this form of memory does not last for more than a few days. Plasmodia do not have any longer-term internal memory. They are troubled by the slime of their childhood, but soon forget an afternoon attack by a beetle that damaged some of the veins.

Through a wealth of experiments on slime moulds, we have seen that they are sensitive to their surroundings, display intelligence as they search for food, learn from their successes and failures, and make decisions about where to travel next. (Danger is represented in experiments by light exposure, which is effective because slime moulds prefer shade.) The giant amoeba does all this with no more brain than Scarecrow in *The Wizard of Oz*. One of the consequences of being brainless is the loss of consciousness. Slime moulds possess intelligence without consciousness; they are aware without being conscious of their awareness, alert without knowing that they are alert. This seems a significant departure from what it is like to be a bat, a whale or a human. Unlike slime moulds, we become mindful of passing seconds when we pay attention to them. But when we inspect the time frames on either side of seconds, the ostensible shallowness of the slime mould recedes from view. Like the lives of animals, the existence of slime moulds relies on a continuous whirl of biochemical routines that take place within fractions of a second. Our fundamental cellular mechanisms are identical. On the other side of the seconds time frame, plasmodial behaviour, like ours, becomes apparent over minutes and hours, but none of us can fix our continuous attention over these longer intervals. We are conscious that the last hour has passed, but we lived most of it in a somewhat distracted fashion, without tracking every moment. The conscious human experience happens in seconds, and slime moulds have nothing like it; the rest of the business of life is not so different.

One final attribute of slime moulds may help to persuade us that Descartes was mistaken when he privileged the human mind above all else in creation. Decision-making that appears to go beyond the rules of a simple programme is one of the keys that has been presumed to set us apart from robots. Until recently, it was fashionable to regard insects as unthinking machines, but ethologists and neurobiologists have triumphed over this fiction. Even without a brain, the slime mould seems smarter than a robot. Humans presented with two bottles of wine that vary slightly in quality and price will choose to purchase either bottle with roughly equal frequency. The addition of a very cheap, low-quality wine to the options has a profound effect on the consumer, increasing the likelihood that they will choose the cheaper of the two original bottles. In the second experiment, nothing has changed in the price or quality of the original bottles, but the decoy introduces noise into the decision-making process, making the consumer rank price above quality. Slime moulds do exactly the same thing.[17] Without a decoy, plasmodia will choose a high-quality food item (attractive) offered in the light (unattractive), and a lower-quality food item (unattractive) in the dark (attractive), with equal frequency. When a very low-quality food decoy is offered in the dark, the plasmodia tend to ignore the high-quality food in the light and increase their preference for the lower-quality food in the dark. They appear to be as irrational as shoppers in a supermarket.

Deep down, of course, slime moulds and humans are both behaving deterministically, responding to programming that is responsive to changes in environmental circumstances. Neither organism has free will. One of the few fundamental differences between animals and slime moulds is boredom. Slime moulds do not suffer from it. Educational psychologists have suggested that the attention span of students is limited to between ten and fifteen minutes, and that lectures should be limited to this time frame.[18] However, this conclusion rested on flawed experiments in the 1970s, which showed a reduction in note-taking as a lecture proceeded. More recent studies on our attention to web pages have indicated that the human attention span is only eight seconds. Goldfish can maintain their focus for at least a second longer than that. Other studies have suggested that human attention is dwindling from generation to generation. But there is very little merit in any of this, because we have no idea how to measure attention.

The idea that attention is an endangered resource, and that computer use is to blame, is at odds with the observation that some video-gaming enthusiasts become so engrossed with the sport that they succumb to severe dehydration and collapse from exhaustion. If a lecture is boring, concentration for no more than a few minutes is understandable. The same goes for books and films, music, sports and any other entertainment that we absorb over minutes and hours. Time passes minute by minute and hour by hour without continuous scrutiny of the accumulating seconds. This slippage is evident when we look at the clock and think about what happened, and what to do next. And always, time flies while we attend to the immediate tastes, smells, comforts and discomforts, the he said and she said, and the meowing of the cat. We live in seconds and behave over minutes and hours.

So much for the busiest forms of animal behaviour. We turn next to plants, whose twists and turns are dictated by the more leisurely tempo of Earth's quiet rotations about itself and around the Sun.

FOUR

BLOSSOMS
Days, Weeks and Months (10⁰–10⁶ Seconds)

Every silent rotation of Earth is measured by species that bathe in the energy of the Sun, some quickening to the rising star, others turning to sleep, as the whole chattering zoo and leafy garden is drawn into another day. In this place, 8 light minutes from the solar furnace, biology has come along for the ride by adopting cycles of behaviour that play out over Earth-days, and the weeks and months of our manufactured calendar. Plants, animals, fungi and the microbes that outnumber them have been shaped to travel like bobsleds, sweeping and turning along their tracks, speeding and slowing to the rhythm of the day, ferrying their genes into the future.

The presence of a timekeeper in plants is evident from the opening and closing of waterlily flowers and the release of perfumes from orchids to attract pollinators. Leaf movements are another example of timed movements, with beans and wood sorrel drooping their solar panels at dusk and raising them to greet the Sun. These expressions of plant behaviour are circadian in nature, which means that they repeat every 24 hours. Time is measured inside the leaf cells from the ebb and flow of 'clock proteins' that switch one another on and off, equipping the plant with an adjustable sundial.[1] The fine-tuning comes from changes in day-length that shift the rhythms of the proteins, but the clock keeps running pretty well even when botanists keep plants in the dark or expose them to continuous light. This resilience signifies the operation of an internal mechanism rather than a fresh response to each sunrise and sunset. Chemical timepieces are used throughout nature, and varying numbers of interactions between clock proteins allow bacteria, fungi and animals to register the days of their lives.

The 'father of botany', Theophrastus, recounted the leaf movements of the tamarind tree in his *Enquiry into Plants*.[2] He drew his description of the daily opening and closing of the leaflets from the observations of an admiral of Alexander the Great on the island of Tylos, now Bahrain, in the fourth century BC.

Water lilies, *Nymphaea* species, whose flowers open and close according to a circadian rhythm.

The people of Tylos, wrote Theophrastus, 'say it [the tamarind] goes to sleep'. Leaf movements appealed to Percy Bysshe Shelley, whose animated vegetable was the protagonist of his poem *The Sensitive Plant*, published in 1820:

> A sensitive plant in a garden grew,
> And the young winds fed it with silver dew,
> And it opened its fan-like leaves to the light,
> And closed them beneath the kisses of night.[3]

Shelley's poem follows the sensitive plant and the other flowers in his garden from their springtime reawakening to autumnal collapse, when 'agarics, and fungi, with mildew and mould / Started like mist from the wet ground cold'. At the end of the poem, he questions whether the sensitive plant, 'Ere its outward form had known decay', felt the return of spring in the following year. Shelley twisted the straight arrow of time into a gyre that traced the death and rebirth of the garden with each turn, far into the future.

The common name of 'sensitive plant' describes *Mimosa pudica*, a legume that grows as a weed across the tropics. It is possible that Shelley thought of this species when he composed his poem, but it does not fit his description very well. *Mimosa* adheres to the circadian cycle of leaflet opening and closing described in the opening lines, but is much better known for rapidly closing its leaflets from one end of the leaf to the other when it is disturbed.[4] This swift motion may deceive herbivores – 'Nothing worth eating here, you were mistaken, try over there' – and exposes a backup deterrent in the form of curved thorns along its stems and branches to animals that persist in their advances.[5] The slower circadian movements of the leaves of *Mimosa*, tamarind trees and other plants cannot be related to herbivore activity because they work in the wrong direction (opening in the morning) to discourage animals that feed in the daytime. It is possible that leaf closure at night drives the whole routine, and plausible explanations for this movement include the avoidance of frost damage, shedding water droplets and reducing the number of infectious fungal spores that land on the leaf surface. None of these solutions is very satisfactory, yet the plants keep at it, opening in the morning light and closing to the stars.

Plant movements may have encouraged superstitions about vegetable spirits. Almost every culture has them, and human transformation into trees appears as a frequent curse in mythology. There are twenty or more examples of these 'mutations' in Ovid's *Metamorphoses*, Virgil turns Polydorus into a myrtle bush in the *Aeneid*, and Spenser figures a tree that bleeds and cries with pain in the *Faerie Queene*. Thomas Vaux, a sixteenth-century nobleman and companion of King Henry VIII, described the suffering plant in his heartbreaking poem 'No Pleasure Without Some Pain'. This is the first of its three verses:

> How can the tree but waste and wither away
> That hath not sometime comfort of the sun?
> How can that flower but fade and soon decay
> That always is with dark clouds over-run?
> Is this a life? Nay, death you may it call,
> That feels each pain and knows no joy at all.[6]

All movements of leaves, whether they happen quickly or slowly, are caused by pressure changes in their stalks. Prayer plants and beans show circadian movements that are inconspicuous from minute to minute, but arresting in time-lapse videos that compress hours into seconds. These species are equipped with a thickening at the base of the flattened blade of their leaves that acts as a joint. This is called the pulvinus. When cells on the lower side of the pulvinus are swollen with water, the leaves are held outwards to face the overhead sun. When the same cells lose pressure and shrink, the leaves turn down towards the stem. It is more complicated than this, actually, because one side of the pulvinus swells as the other shrinks to raise the leaves, and the reverse happens to lower them. This action resembles the reciprocity of extensor and flexor muscles in limb movements. Circadian leaf movements rely on changes in the levels of clock proteins. These contrast with rapid leaf movements that are triggered when the plant detects that its tissues are being compressed. In both cases, the signal for pressure changes is passed by electrical signals from cell to cell. At the cellular level, plant sensitivity is based on mechanisms that bear comparison to the transmission of nerve impulses in animals.

Even the least animated plants throb with life when they are not shut down by frigid weather or baked dry by the Sun. As Earth turns, plants close their microscopic stomata or breathing pores during the day to conserve water, and ramp up levels of defensive chemicals to rebuff fungal pathogens that initiate their infection mechanisms according to their own circadian programming. Plants are continuously adjusting the position of their leaves, and the movement of water through the xylem tissue and syrupy sap in the opposite direction through the phloem causes periodic changes in the girth of the plant. There is a tendency to treat these oscillations as emblems of animalistic sensitivity. Investigators in the 1960s took this to an extreme when they claimed that plants became relaxed when they 'listened' to Beethoven, and panicked when they were threatened with a burning match. These clowns were gullible at best, foolish in their dismissal of control experiments and besotted with fantasies about the inner emotional lives of the botanical world.[7] Even the most enthusiastic botanists must admit that, lacking a brain, plants are limited in their apprehension of the environment compared with an animal equipped with a knob of neurons. While the fundamental cellular mechanisms of sensitivity are the same, a squirrel is greatly distressed when a hawk strikes and carries one of her pups into the sky, whereas the plant is forever indifferent to the squirrel eating its seeds.

Simpler than any plant, pigmented bacteria that carry out photosynthesis are also responsive to the rotation of the Earth, measuring time with their own very streamlined sets of clock proteins. Groups of these molecules are arranged in the shape of pairs of stacked doughnuts and undergo a sequence of chemical modifications during daylight, which are reversed overnight. This intricate timepiece is wound and unwound, day in and day out, like the mainspring of a mechanical watch. It is reset by sunlight, but, unlike the plant clock, it loses track of time quite quickly when the bacteria are kept under conditions of unremitting light or darkness. Despite this limitation, the attention paid by bacteria to the planet's spin cycle is vital because it allows them to harmonize the expression of different genes, and the cellular processes controlled by these genes, with the availability of solar power. During the night the bacteria draw on energy reserves to chug along, and by anticipating sunrise they 'know' when their batteries will be recharged and can prepare for the day's work.

Other kinds of bacteria that consume organic compounds in their environment, rather than making their own food by photosynthesis, engage in circadian rhythms without responding to sunlight. Think about the trillions of bacteria in the human gut that break down the more complex plant materials in our diet into simpler compounds that we can metabolize. These microbes live in perpetual darkness, but their rhythms are tied to the Sun because most of us suspend our overnight fast and begin snuffling around in the kitchen around daybreak. In a satisfying twist to the ostensible master–slave relationship at work here, the bacteria exert some control over our eating habits, when they inform us that *they* would like a snack by stimulating our appetite.[8]

Plenty of other organisms 'hath not sometime comfort of the sun'. These include the residents of the dark biosphere that blossom in the blackness of caves and in the depths of the soil or the sea, in marine muds and around hydrothermal vents. Circadian rhythms would seem superfluous for organisms that grow so far from the light, but there is evidence of daily behavioural rhythms among some of the animals associated with the sulphurous dark smoker vents of submarine volcanoes. Observations made from unmanned platforms attached to the Pacific sea floor at a depth of 2 km (1¼ mi.) have revealed that the iconic red-plumed tubeworms play peekaboo from their crystalline tubes with some synchronicity, and that the maximum number of exposures occurs every 12 or 24 hours.

Other kinds of worm that browse the surface of the vent chimneys show more frequent rhythms of activity, but there do not seem to be any repetitive changes in the deportment of the sea spiders and crabs that crawl on the outside of the chimneys. The sea spiders bounce 'up and down, bending their legs, sometimes on top of another individual', according to their spidery whims rather than at particular times in the unremitting dark.[9] And to think they have been dancing, silently and shadowless, for hundreds of millions of years along cracks in the sea floor where fountains of scalding mineral water disgorge into the abyss. Milton's description of hell is perfect for this sea-spider lair: 'As one great furnace flamed, yet from those flames / No light, but rather darkness visible.'[10]

In the absence of dawns and dusks it is not obvious how anything can tell the time, and that makes it difficult to explain the diurnal cycle of tubeworm behaviour. Clues to a possible mechanism come from videos recorded from a pair of deep-sea observatories in the Atlantic and Pacific.[11] These show that the tubeworms pop out six hours earlier in one location than the other, which corresponds to the difference in the time zones occupied by these isolated vent communities. The explanation that follows from this time lag is that the rhythms of the worms may be driven by changes in water temperature and oxygen levels, and that these are linked to the tidal pull of the Moon.

Our satellite affects life on land, too, through its monthly changes in luminosity. Once a year, corals release their eggs and sperm cells at night in synchronized mass spawning events that follow the brightness of a full moon. The full moon also stimulates salamanders and other amphibians that migrate to ponds and seasonal pools to engage in orgies of explosive breeding; the nesting cycles of nightjars obey the lunar cycle, too, and some mammals are friskiest when the 'Lozenge of love! Medallion of art!' – Philip Larkin's mocking description in 'Sad Steps' – brightens us with the whole of its brilliant hemisphere.[12] The excavations of the antlions described in Chapter Three are another instance of life taking its cues from moonlight, with larvae digging larger pit traps under a full moon. The reason for this behaviour is not known, but the insects must have an internal lunar clock, because they continue to dig wider and deeper cones every 28 days even when they are reared indoors. Like werewolves, they have no choice but to follow the call.[13]

Organisms that live in caves, called troglobites, compete with the strangeness of the life that has evolved around hydrothermal vents. Cave worms and subterranean shrimp, spiders, insects, fish, salamanders and other animals have evolved from surface-dwelling ancestors. They have lost their eyes and pigmentation over millions of generations, and have abandoned the daily see-saw of surface temperature in favour of the constant climate of their grottos. The energy that is used to grow eyes is conserved in these cave species, which have suppressed the genes that work to construct eyeballs in their relatives on the surface. These genes are not scrubbed from the genome, however, and cave fish begin the process of crafting eyes in embryo before allowing them to degenerate.[14] Without eyes, these animals do not respond to light exposure, but they retain parts of the chemical clock mechanism of their ancestors and use it to mark time when this is advantageous. This capability is apparent in cave fish that become more active in readiness for their regular feeding time after a period of training by researchers.

Circadian rhythms play such an important part in our lives that we take them for granted until they are disturbed by putting the clock forward in the spring and backwards in the autumn, or, more profoundly, by flights across several time zones. As a deep sleeper with an unusually theatrical dream life, I have a body clock that runs with the precision of a Swiss timepiece until it is transported westward to Asia. Even after a week in Japan I cannot sink into unconsciousness for more than a couple of hours a day, and am forced to enjoy the splendours of this country as a round-the-clock zombie. Back home, Morpheus is always close at hand, leading me to a minimum of eight hours sleep each night, and stopping by to augment the afternoon with a nap whenever permissible.

The human body clock is similar to the plant timer, with component genes that turn each other on and off in tune with the appearance and disappearance of the Sun. It is programmed by 20,000 nerve cells that form a structure called the suprachiasmatic nucleus in the hypothalamus. The hypothalamus sits on the underside of the brain and is the size of an almond. Nerve cells in the retina transmit information about light levels to it, and it responds by directing melatonin production by the pineal gland. Melatonin acts as a sedative by slowing brain function, and more of it is produced in low light. Jet lag exposes the stubbornness of the molecular clock, which, entrained in Ohio, tells me to be wide awake even when my befuddled brain can see that it is the dead of night in Tokyo.

The pineal gland is a standout in the evolution of the vertebrate clock. It is elaborated into an organ called the parietal eye in some reptiles, amphibians and fish. In the lizard-like tuatara of New Zealand, this third eye sits between the lateral eyes, on top of the head. It is fitted with a tiny lens that focuses light on a cup of light-sensitive cells that functions as a retina. Parietal eyes are simpler in other reptiles and covered by skin, but in all species they seem to function in setting the circadian rhythm. The human pineal is no larger than a grain of rice, but René Descartes regarded it as the *siège de l'âme*, the seat of the soul, the anatomical location where the thinking mind interacted with the mechanical body.[15] Part of the logic for his proposal rested on the unpaired nature of the pineal, which contrasted with paired eyes, ears and cerebral hemispheres. The philosopher regarded this as a sign that it could direct 'animal spirits' through the brain so that, for example, we do not see two images of a single object. He examined the pineal glands in cow brains purchased from slaughterhouses in Amsterdam, and thought that although these fleshy pips served as information processors, their activities did not rise to the level of the exclusively human soul.

Descartes had unusual sleep patterns. According to his biographer Adrien Baillet, the philosopher went to bed late at night and slept ten hours or more for most of his adult life. He experienced 'enchanting dreams of trees, gardens, and palaces ... [and awoke] to more complete contentment'.[16] This could qualify as hypersomnia, which is a recognized sleep disorder, but 'continuous fulfilment syndrome' would seem a more appropriate diagnosis. It is the polar opposite of the horror of fatal familial insomnia, which is caused by a mutated gene that spawns a misfolded prion protein that destroys masses of brain cells and prevents the patient from sleeping at all. Descartes was forced to abandon his luxurious sleep pattern at the end of his life, when he was tutor to Queen Kristina of Sweden, who preferred her lessons in the early morning. It has been suggested that this perturbation of a lifetime sleep pattern may have led to his death from a viral infection, at the age of 53, by weakening his immune defences. There is some irony here, in the privileged place Descartes accorded to the tiny organ whose melatonin secretions he was compelled to overcome to awaken at dawn in his freezing bedchamber in Stockholm.

All animals sleep in some manner, whether they fall unconscious for a good chunk of each day or just cycle between periods of greater and lesser activity as the planet turns. Jellyfish have such simple nervous systems that the idea that they sleep was dismissed until it was found that they reduce the pulse rate of their bells at night. We can rouse them from this listlessness by disturbing their water or offering food, but repeated sleep deprivation results in less activity during the following day. More surprising still for species lacking a brain, jellyfish respond to a dose of melatonin by going to sleep. Nematode worms and fruit flies are also entrained to circadian patterns of sleep, which begs the question: why do animals sleep?

The most compelling explanation is that daily periods of physical inactivity are inevitable on a planet that exposes its residents to alternating light and darkness.[17] Sleep conveys the dual advantage of conserving energy and reducing the risk of being eaten. The little brown bat, which we considered in Chapter Three, sleeps for up to twenty hours per day. By launching themselves at their insect prey at dusk, they consume enough calories to meet their immediate needs and to store enough fat reserves for reproduction. What else is there to being a bat? Muscular exercise in daylight would serve no purpose, and flying during the day would be hazardous in an environment populated by sharp-eyed birds of prey. The echolocating skill of little brown bats and the circadian rhythms of their insect targets are attuned to darkness. The same sort of adaptive explanation works for other animals. Sleep performs other functions, such as tissue repair, which explains why we feel rested physically after a good night's sleep, and the foggy state in which we emerge from a sleepless night shows that we also need sleep to refresh our computers.

The emotional functions of sleeping must be an evolutionary add-on to an original mechanism, because it seems unlikely that nematode worms and jellyfish need to revisit their worries and desires while they snooze.

For animals with complex brains, however, sleep is accompanied by dreams. In his essay 'The Terrors of the Night', Thomas Nashe wrote: 'A dreame is nothing els but a bubling scum or froath of the fancie, which the day hath left undigested.'[18] This Elizabethan notion is very close to the current view of dreaming as a mechanism for sorting through the mass of superfluous information gleaned from the day, and archiving the important stuff for future access. Memory processing and consolidation are widespread functions of sleep in other animals. Without adequate sleep, honeybees lose their navigation skills, butterflies lay their eggs on the wrong plants and young fruit flies develop lifelong learning problems. Day in, day out, we dream, whether we walk with Descartes in perfumed gardens or are shoved by a leather-aproned monster into a torture chamber. The rarest and most rewarding dream comes when the imagination conjures a tale of such lucidity and lunacy that we are awoken to discover ourselves crying with laughter.[19]

The extension of life over weeks and months is an exercise in the repetition of circadian rhythms. On Earth's surface, these diurnal patterns are adjusted to match seasonal variations in food availability and to meet the needs of individual reproductive cycles. Land plants and aquatic algae respond to temperature and day-length, pressing their greenery upon great swathes of the planet. As springtime spreads across each hemisphere, botany explodes over the land and the oceans become cloudy with algae. Bamboo species, which are grasses, are among the fastest-growing plants, with rates of stem elongation approaching 1 m (3 ft) per day. This growth is audible as a popping sound in a grove of bamboo plants, as the tissues wrench themselves apart. Giant kelp, which can extend its fronds 45 m (150 ft) in a single growing season, is the fastest-growing seaweed. Wherever there is liquid water, the planet pulses with life. Unaware of this wider global rhythm, we follow the dictates of our 24-hour body clocks, go about the immediate business of existence and are conscious, intermittently, of the flow of seconds in the pursuit of tasks that occupy minutes and hours. As these shorter intervals are pushed into history, we notice the change in day-length, the Sun rising earlier each morning as the weeks and months glide by, and then later again, until too soon we are surprised to find ourselves in a new year.

In his poem 'My Descendants', W. B. Yeats wrote: 'The Primum Mobile that fashioned us / Has made the very owls in circles move.'[20] For me, this speaks to the way that all life on this Goldilocks planet dances to the light and the darkness of its spin. Circadian rhythms are also embedded in the lovely poem 'The Water Mill' by Fredegond Shove (the poet from whom I quoted my opening epigraph to the Preface). Here are a few lines to close this chapter:

> The miller's cat is a tabby, she
> Is as lean as a healthy cat can be,
> She plays in the loft where the sunbeams stroke
> The sacks' fat backs, and the beetles choke
> In the floury dust. The wheel goes round
> And the miller's wife sleeps fast and sound.[21]

FIVE

BROODS

Years (10^7 Seconds)

After an extended infancy and years passed as sullen and uncommunicative adolescents, cicadas and humans are reborn as bright-eyed adults, ready to take on the world and eager to sing of love. Our childhoods are extended to the same length as the thirteen- or seventeen-year cicadas of North America, which are among the longest-lived insects. After all that time spent in darkness as wingless youngsters, or nymphs, the cicadas enjoy a brief summer romance of singing and mating, singing and mating. The Roman poet Virgil celebrated 'the plaintive cicadas [who] thrill the orchards with song'.[1] The Greeks loved cicadas, too. Homer likened their sound to the old men of Troy in the *Iliad*, and Socrates tells a tale of human transformation into cicadas in Plato's *Phaedrus*. Daoists refer to the moulting process of the adult cicada as an act of *shijie*, or disengagement from the corpse, which confers immortality on the insect. Cicadas have been associated with death and resurrection for millennia. Beautiful jade carvings of life-sized insects from the Chinese Han dynasty (206 BC–AD 220), when Daoism or Taoism (same thing) took shape, are found in museum collections and are popular at auction. These amulets were slipped on to the tongues of the deceased, or bored with little holes to be worn as pendants by the quick.[2]

The synchronized emergence of billions of 'periodical' cicadas every thirteen or seventeen years is a distinctly North American spectacle. Entomologists have described 3,000 species of cicada, mostly from the tropics. They are classified as hemipterans, or true bugs. The periodical cicadas are classified as species of *Magicicada*, which is a name that evokes the seeming sorcery of their appearance. The great majority of cicada species emerge from the soil after a few years, without synchronizing their appearance to produce recognizable broods. These are the annual cicadas that keep buzzing every year, creating the melodious back-drop to the Tuscan luncheon of my dreams beneath ancient olive trees, around a table set with carafes of wine and baskets of bread on white linen. (Lunch today was a dish of microwaved potatoes and Brussels sprouts with the online edition of the *Washington Post* for company.)

The standard explanation for the monstrous outpouring of the periodical species is known as the 'predator satiation strategy'. By crawling from their burrows and reaching maximum density in a single day, the cicadas overwhelm their predators. Birds and mammals, including raccoons and opossums, gorge themselves on the crunchy-on-the-outside, gooey-on-the-inside insects with-out denting the brood. (The description is based on a regrettable personal tasting and informed my guess about the experience of a bat eating a moth in Chapter Three.) This ensures that a superabundance of nymphs plant themselves in the soil before the end of the summer. Mayflies profit by the same tactic, form-ing gargantuan summer swarms of adults, whose sexual frenzies are detected as clouds on weather radar in the Upper Midwest of the United States. The precise timing of the mayfly extrava-ganza is more critical than the materialization of the periodical

Periodical cicada, *Magicicada cassinii*, whose broods emerge in
North America on a seventeen-year cycle.

cicadas, because the smaller insects die within hours or minutes after they moult from their larval phase and take wing. Females of one of the less common mayfly species, *Dolania americana*, mate and deposit their eggs in less than five minutes; they are the shortest-lived insects, while the males patrol their streams for a little longer before dropping into the water from exhaustion.[3]

The prime-number extension of the cicada life cycle is puzzling. It may be adaptive, in the sense that it conveys some advantage, or it could belong to a quirk of evolutionary history from which the insects gained no special edge over their competitors. According to the adaptive model, a thirteen- or seventeen-year cycle prevents predators from matching their life cycles to the spikes in insect numbers. Over a decade, cicadas that emerge every two, four and six years will be vulnerable to predators that emerge every one, two, three, four or six years, and the two-year cicadas will also fall prey to a five-year predator in year ten. If multiple broods of cicadas cycled on these schedules in the same region, their predators would always have something to eat. Extended prime-numbered broods are different. They escape from anything other than a perfectly matched prime-number predator, which could, should it arise, drive itself to extinction by devastating the target brood.[4] In an interesting twist to the predator–prey relationship, populations of the wood-peckers, blue jays, grackles and cardinals that devour the cicadas appear to crash in the years following the prime-number broods. This might be because of the boom in bird reproduction fostered by the feast, and bust in the famine that follows.[5] Whatever the mechanism, the cicadas enjoy some protection from the recession of their enemies on their next emergence.

The adaptive model for the prime-number behaviour seems very neat until we reflect on the fact that the annual cicadas and their predators enjoy balanced relationships that allow both to travel through time. An alternative, non-adaptive model for the periodical cicada requires us to think about the weather conditions in North America in the Pleistocene Epoch, or Ice Age. Soil temperatures would have been very low during glacial periods, and that would have extended the juvenile stages of the refrigerated insects over many years.[6] Over tens of thousands of years of fluctuating temperatures, broods would have emerged after prime and non-prime spells in the ground. Cicadas whose life cycles were drawn out over thirteen or seventeen years were less likely to mate with insects on other emergence schedules, which had the effect of stabilizing their prime-number programming.

Either way, the explosion of cicada adults is an awesome business. For me returning to Ohio after a trip to Britain in 2004, the sudden exposure to the cacophony produced by Brood X verged on the hallucinogenic. It came in waves of unreasonable volume from the tree canopy, with brief intermissions before the insects roused one another like football fans to launch the next shimmering chorus. Cicadas are the loudest insects. In 1633 William Bradford, the governor of Plymouth Colony, described their 'constant yelling noise . . . [which] made all the woods ring of them, and ready to deaf the hearers', and, mistaking them for locusts, the eighteenth-century naturalist Paul Dudley quoted the Book of Joel (2:5), in which they are described as making 'the noise of chariots on the tops of mountains'.[7] The males sing by buckling a pair of abdominal 'tymbal' membranes hundreds of times per second, achieving something akin to the 'thunder sheets' used on stage for the ferocious storm on the heath in Shakespeare's *King Lear* (Act III, Scene ii):

And thou, all-shaking thunder,
Smite flat the thick rotundity o' the world!
Crack nature's moulds, an germens spill at once,
That make ingrateful man!

After mating, the silent females cut a series of slits in the bark of tree branches using a saw-toothed ovipositor and lay their eggs in the wounds. Six weeks later, the nymphs hatch from these 'egg nests', drop to the ground, bury themselves in the soil and begin feeding on tree roots. Analysis of tree rings shows that periodical cicadas reduce tree growth by as much as 30 per cent. Despite the damage, particularly to fruit orchards, the major broods should be seen as a blessing. They are one of wild nature's last stands on a planet that has lost its greatest zoological extravaganzas. The flocks of passenger pigeons that blackened the Midwestern skies vanished in the nineteenth century, and the last herds of bison that migrated across the Ohio River were extinguished in the previous century, but, before the rest of biology goes silent, we are left with the spectacle of a tonne or more of shrieking cicadas per acre (0.4 ha) of forest.

If she had never moved from southwestern Ohio, an elderly resident born on a farm in 1900 would have heard Brood X six times: first, as a two-year-old in 1902, the year before the Wright brothers from Dayton took to the air; next in 1919, when her boyfriend returned from France; and three more times before her last serenade in 1987. The cicadas were underground as most of her life passed, exposed only when she spaded through their chambers to plant her tomatoes, disturbing their doze in the dark with the brilliance of the Sun and making them squirm like vampires. The call of Brood X is an incantation of mortality – theirs and ours. Like the toll of John Donne's bell, 'do not ask . . . it tolls for thee.' The children of the brood that departed in 2004 will sing to me in 2021, but 2038 seems a bridge too far, for me, and perhaps also for the insects in this warming world. In the meantime they deserve the benediction offered by an unidentified Greek poet in the *Anacreontea*: 'We count you blessed cicada, when on the treetops, having drunk a little dew, you sing like a king: you own everything that you see in the fields, everything that the woods produce.'[8]

The human reproductive cycle can certainly be completed in the same timespan as a seventeen-year cicada, although the fact that young women reach peak fertility in their early twenties suggests that evolution has graciously favoured a lifespan that extends as far as our third decade. Part of the reason we cannot get everything done more swiftly is that human infants remain so desperately needy for so long. Mouse gestation takes 19 to 21 days. Pups are born pink, their eyelids and ears closed, and feed on their mother's milk for the next three weeks. Female mice reach sexual maturity when they are six weeks old, bear between five and ten litters of pups per annum, and can live for two or three years if they elude their predators. At the other extreme, sperm whale mothers cultivate their foetuses for up to twenty months and African elephants are pregnant for six weeks longer than that. Within minutes of their births, however, whale and elephant calves can swim and walk on their own, while we have difficulty keeping upright after a year of gurgling and continue to grab chair legs for support.

Some biologists suggest that this protracted delicacy results from the time it takes to forge such an intellectual giant. If mothers were capable of carrying us for longer in the womb, say eighteen months, we would be born with a bigger brain, shorten our period of larval squirming, and get on with the serious business of life like other animals.[9] Mothers interviewed for this chapter were, understandably, angered by the thought experiment of doubling the length of a pregnancy. Putting aside their feelings, the 'obstetrical dilemma' hypothesis proposes that evolution has not permitted an eighteen-month pregnancy because the adjustment of the pelvic bones to accommodate a wider birth canal would interfere with walking upright. This meant that as brain size increased in our hominid ancestors, it became essential for mothers to give birth to immature babies that developed their elementary life skills during an 'extra-uterine spring'. The validity of this idea is weakened by biomechanical studies demonstrating that a broader pelvis does not limit the efficiency of the female body as a machine for walking and running.[10] This suggests that women could, in theory, carry a larger foetus with further evolutionary modification of the pelvis.

An alternative reason that we are born in such a sorrowful state may be that our mothers cannot satisfy the energy needs of the foetus any longer than nine months.[11] Day after day, a pregnant woman approaches a level of energy expenditure associated with long-distance running or cycling, pushing her body to the limits of physical endurance. Indeed, humans already hang on to the foetus for more than a month longer than we would predict by looking at other primates of similar body mass. We expel the baby when it becomes too stressful to keep feeding the parasitic foetus through the placenta and switch to breastfeeding as the nutrient-delivery system for the baby. Breast milk is the perfect food, rich in short-chain fatty acids that are essential for brain development.

The monthly oestrous cycle in humans is comparable to the periodicity of ovulation in other great apes. There is a weak seasonal effect on the rate of conception in humans, which leads to an increase in spring births in the northern hemisphere, but we lack any verifiable annual reproductive rhythm. This is convenient for astrologers, who would be hard put to create horoscopes if we all shared the same star sign, as frogs do. Seasonal effects on other features of human physiology also seem to be relatively mild, and the comforts of the modern world buffer many of us from having to do much more than add or subtract layers of clothing in response to the weather forecast. In contrast to this scant regard for the progression of each year, many other animals and plants are tethered to the annual circuit around the Sun and behave according to strict rhythms of internal 'circannual' physiology.

Dinosaurs followed circannual cycles in their mating and egg-laying behaviour. Fossils of the Ornithopoda, the bird-footed dinosaurs, reveal how these extinct reptiles responded to annual weather patterns in Africa during the Triassic by entering a period of dormancy during the dry summer months, and mating and laying eggs in the rainy season that followed.[12] The ornithopods include the famous Iguanodon and duck-billed hadrosaurs. Until recently, palaeontologists thought herds of American hadrosaurs migrated southwards from the Alaskan North Slope to winter feeding grounds, as caribou do today. More recent evidence suggests that at least some of these animals stayed put and must have been adapted to cold weather. Evidence for seasonal migration is stronger for giant sauropod dinosaurs whose fossilized teeth have different isotope signatures from the Jurassic rocks in which they are embedded.[13] Analysis of tooth enamel from species of

Camarasaurus show that the animals left the summer heat of the Wyoming lowlands 150 million years ago and walked 300 km (185 mi.) to the cooler highlands of Utah. These long-necked giants browsed on the upland vegetation during the summer before making the return journey to the river basin in Wyoming where they overwintered.

Aerial migration over much greater distances may have been mastered by pterosaurs, which were capable of gliding for thousands of kilometres, and shorter annual journeys may have developed in early birds. The migratory birds of today are so punctual that their comings and goings are embedded in our culture. In Britain, the first call of the cuckoo, *Cuculus canorus*, after its winter sojourn in Central Africa is the harbinger of spring. The nightingale, *Luscinia megarhynchos*, arrives from Africa at about the same time, and these tuneful species have competed for our affection for centuries. Following ancient tradition, John Milton contrasted the nightingale, 'That on yon bloomy spray / Warblest at eve, when all the woods are still', as good omen for the poet in love, with the song of the cuckoo, 'the rude bird of hate', which foretold his 'hopeless doom'.[14] I used to hear the nightingales in an inner-city park in Bristol in my student days, and can conjure their bewitching note even now. Rachel Carson wrote about the 'symbolic as well as actual beauty in the migration of the birds', and found 'something infinitely healing in these repeated refrains of nature'.[15] Unfortunately, there are few places where nightingales can be heard in Britain anymore.

Proof that some circannual cycles are controlled by clock mechanisms that are built into the structure of animals comes from the behaviour of captive songbirds. Migratory birds kept in cages become restless, hopping about, fluttering between their perches and flying at night, at the same time that unrestrained members of their species set off on their journeys north and south. The German word for this behaviour among birds imprisoned in the name of science is *Zugunruhe*. Other indications of the circannual clock come from the reproductive behaviour and moulting rhythms of birds kept at a constant temperature and uniform light-dark cycles. Species subjected to these torments include willow warblers, wading birds and an African bird called the stonechat that has maintained its circannual behaviour for more than a decade in captivity.[16] Proving that the behaviour is instinctive, stonechats hatched in Germany replaced their feathers on the same schedule as their Kenyan parents who had been captured in the wild.

Even when a timer is ticking away inside an animal, it is likely to be responsive to changes in day-length and temperature. Unlike atomic clocks, biological timers are receptive to the environment. This flexibility is crucial for preventing a rodent from freezing to death if its alarm clock is ringing, 'hibernation over', and its burrow is buried beneath deep snow from a late storm. We know, for example, that chorus frogs in North America, called spring peepers, are programmed to start chirping and mating after winter hibernation, but they will stay silent until the temperature rises and rain showers flood the shallow pools where they lay their eggs. Changes in hormone levels have been measured in frogs kept at a constant temperature, a fact that certainly points to an intrinsic clock. The spring migration of huge numbers of spotted salamanders to their mating pools is another example of the interplay between environmental conditions and a physiological clock. The reproductive organs of these animals have been ripening for weeks during hibernation according to a *circannual* rhythm, and on the first warm night in spring under gentle rainfall, a *circadian* clock stimulates the slippery march to the mating pools under cover of darkness.

It is very unlikely that any organism possesses a circannual molecular clock that resembles the structure of the protein-based timepieces that track the 24-hour cycle in animals, plants, fungi and microbes. Instead, circannual behaviour is based on the crosstalk between various physiological processes that are linked to circadian rhythms and to the rates of a variety of developmental processes. The increase in day-length in the weeks before spring may trigger the characteristic increase in testis size in male salamanders and corresponding changes in the activity of the ovaries of the females. These developmental mechanisms may be responsive to temperature, but they cannot be accelerated beyond a certain point, which places constraints on the mating schedule. When the animals are on the move, beneath a steady April shower, they are responding to the interplay between a series of timing mechanisms within their bodies and the likelihood that they will find pools of water that have been filling with water at around the same time in April for hundreds or thousands of years. The animal world is filled with migration rituals that have replayed for millennia, from salmon and sardine runs, to the transcontinental flights of monarch butterflies, to the dust-stirring herds of African game and the Christmas Island red crabs that crawl from their woodland burrows and scuttle in their millions to mate at the seaside.

Similar coordination between clocks and climate applies to the life cycles of plants. It takes a certain number of days for seeds to form after the fertilization of the eggs that are enclosed in the female sex organs of a flower. Following their dispersal, the seeds will remain dormant for a particular time, and germination will occur at a permissive temperature and level of soil moisture. The appearance of a perfect circannual cycle emerges from a combination of genetic programming, developmental limitations and the weather. The fruiting patterns of forest mushrooms are galvanized by the same variables. This is illustrated by the edible St George's mushroom, *Calocybe gambosa*, named for the regularity of its appearance on 23 April, which is the patron saint's day in England.

Seasonal patterns of reproduction in marine invertebrates, corals and seaweeds are also driven by overlapping circadian and circannual rhythms, and sophisticated circannual timing is also evident in some microbes. A single-celled bioluminescent alga spends the winter inside a cyst, buried in mud in the Gulf of Maine. In the spring it uses a pair of undulating flagella to power its way upwards until it reaches the water, where its population explodes. This alga is a kind of dinoflagellate that contains a potent neurotoxin, which means that its spring and summer blooms can cause paralytic shellfish poisoning in people who consume contaminated clams and other bivalves. When sediment cores are stored in the laboratory, the algae maintain their annual migration pattern, showing that the circannual behaviour is part of their cellular make-up.[17] This ingenious species is also equipped with an on-board circadian clock that it uses to regulate flashes of blue light as it cruises in the water at night. Most of nature pays attention to the year in one way or another.

Years seem to pass more swiftly as we get older, and we express surprise at the repetition of markers such as birthdays, graduation ceremonies, religious holidays and changes in day-length that creep up on us with alarming regularity. The apparent acceleration of years has been ascribed to the simple trick of memory, whereby the fifth year of a five-year-old covers one quarter of its prior existence, whereas the fiftieth year is just 2 per cent of the quinquagenarian experience. This makes some sense if we consider that a toddler encounters more novelty per year than does an adult who has 'seen everything'. Another reason why a year seems to pass more slowly when we are young is that we may spend more time daydreaming, gazing from a car window, for example, on a seemingly endless trip. The days of my childhood yawned wide in the winter and flashed by in the summer, when the hours were consumed by swinging on a pendulum of knotted rope over the River Thames.

A more scientific solution takes us back to the flicker fusion rate discussed in Chapter Three. This is the neurological mechanism that allows dragonflies to separate two hundred images per second, compared with half that rate in humans. Experiments show that our flicker fusion rate slows over time, which means that the days are stretched in the experience of the child and compressed as we grow older.[18] It follows that years do get shorter if we time them by the quantity of information flow, rather than the unvarying number of seconds it takes to whip around the Sun. If this troubles you for one moment, rest assured that the problem will intensify, and the whole of your journey will be done more swiftly than you ever imagined as a child. My grandmother shared this scrap of wisdom with me in her great age: 'You'll be surprised by how fast it all goes. I was a little girl just the other day and look at me now.' And then she laughed her frog-croaky laugh at the silliness of it all, turning a somewhat maudlin reflection over her afternoon tea into something quite beautiful.

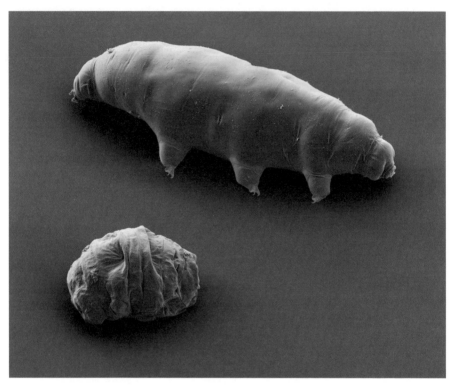

Tardigrade or water bear, *Paramacrobiotus richtersi*, in its active form with legs extended, and in its dried dormant state called a tun.

SIX

BEARS

Decades (10⁸ Seconds)

During my adolescence, I was devoted to an English space rock band called Hawkwind, whose members were in turn devoted to magic mushrooms and sang about interstellar travel.[1] One of their songs concerns an astronaut who has been in a deep freeze for a few decades and awakens to reflect that his earthbound teen girlfriend must be in her sixties now, as he races through the galaxy in a state of youthful preservation.[2] Researchers associated with space agencies have entertained similar fantasies in their enthusiasm for manned and womanned missions within and beyond our solar system. Despite a lot of talk about putting *Homo sapiens* on Mars, and leaving the collapsing biosphere for an unspoiled Goldilocks planet, suspending the animation of astronauts remains the stuff of fiction – and may be so forever. But we have sent water bears into space, which is where this chapter on decades begins.

Water bears, also known as moss piglets, are classified in their own animal phylum called the Tardigrada, which means 'slow steppers.'[3] Water bears are on the edge of microscopic, and the larger ones are visible as pale blobs in bright light. They have eight telescopic legs and blunt snouts, and move like drunken campers trying to escape from their sleeping bags. This clumsiness

93

results from being flooded with water on a microscope slide and dislodged from the surface of the moss plants to which they cling and whose juices they suck for food. Water bears have perfected the trick of anhydrobiosis, which involves drying to a crisp, reducing metabolism to a level that is barely perceptible and persisting in this state of suspended animation for decades. Once they have dried, water bears are exceedingly resilient and are celebrated for their resistance to freezing and boiling, immersion in toxins and irradiation with X-rays. On an unmanned capsule in orbit 260 km (162 mi.) above our planet in 2007, dehydrated tardigrades survived exposure to the vacuum of space and the glare of solar radiation.[4] News reports following this space mission made science-fiction fans giddy with excitement about these 'indestructible' creatures. The actual results were less impressive: only three individuals of one species survived the harshest conditions, and just a single water bear of another species roused itself after punishment by vacuum and radiation. Furthermore, all the survivors died a few days after their brief revitalization.

The urge to hype the brilliance of nature is one of the many less appealing trends in science reportage in our time. Water bears are inspiring animals without any need for fictional decorations. They have been here for hundreds of millions of years, and will outlive us by an eternity. Their ability to cycle through periods of immobility extends their earthbound lives to thirty years, although this feat of longevity is possible only if they have been inactive for most of that time because active water bears live for months rather than decades. The thirty-year record was established by the restoration of a pair of animals from a sample of moss collected in Antarctica. There was a claim in 1948 of a single water bear clawing itself over a moss sample that had been rehydrated after 120 years – 'quivers in several zones of its body were noted' – but the equivocations in the description do not inspire confidence.[5] Water bears suspend their animation in response to desiccation or freezing. Both processes rob the animal of liquid water, a fact that has the effect of slowing the chemical reactions of metabolism to a crawl or bringing them to a complete stop. When the water bear is rehydrated, either with a drop of water or by gentle warming to release water from its frozen state, the engines of its biochemistry roar back into action and the beast is resuscitated.

Animals awakening from this kind of temporary immobility can act like either Sleeping Beauty or Rip Van Winkle. Like Hawkwind's astronaut, Beauty arose fresh as a daisy after decades asleep, whereas Rip Van Winkle found himself an old man with a long white beard following his disappearing act in the Catskill Mountains. Water bears behave according to the Sleeping Beauty model when they are dried and rehydrated, living for days longer than their peers that persisted in their daily tasks without being dried. Freezing the eggs of water bears has the same result, and adults born from thawed eggs show a precise lag in maturation compared with those born from eggs that were cooled. So, whether the animals are dried or frozen, water bears sleep like the princess and awaken biologically younger than their chronological age. Roundworms, or nematodes, can also survive drying, but on revival they are as battered by age as worms that have persisted in their wriggling all the while.[6] The biological ageing of the worm cannot be slowed; they obey the Rip Van Winkle model after a long dry spell. Other animals with the knack for anhydrobiosis include the embryos of brine shrimp, also known as sea monkeys, which live in the Great Salt Lake of Utah; rotifers, or wheel animals; and the larvae of midges.

The tenacity of these invertebrates is not matched in any vertebrates, but some frogs and salamanders are capable of freezing in winter and thawing in spring. The North American wood frog seems to freeze solid, like an ice lolly, or 'ice pop' in American parlance. When any animal tissues are frozen, ice crystals form outside the cell membranes. This concentrates the surrounding fluid, causing the cells to wither by dehydration. As the temperature continues to fall, icing occurs inside the cells, too, which is lethal. Wood frogs survive by promoting the formation of ice in their abdominal cavities and in the fluid beneath the skin, to protect the vital organs. Icing inside the heart and lungs is avoided by the accumulation of sugars and other compounds that lower the freezing point and keep these organs hydrated. Thawing is a stressful process, too, but the heart resumes beating as the ice melts, blood flow is restored and the frog is kicking its legs within hours.

Microorganisms are very good at freezing and thawing, and live bacteria have been isolated from ice cores that are hundreds of thousands of years old.[7] Rather than persisting in a dormant state for this immensity of time, these organisms may survive by continuing to grow and reproduce at a very slow rate. This allows them to tend to the persistent damage to their DNA and keep dividing to produce fresh daughter cells. The maximum age of an individual bacterial cell is unknown. Moving to somewhat larger organisms, Russian biologists claimed to have isolated live nematodes from Pleistocene permafrost that has been frozen for 40,000 years. After several weeks of incubation, they reported adult worms in culture dishes containing thawed samples from permafrost cores. This experiment left lots of questions unanswered.[8]

Cryopreservation of human eggs and sperm cells has been a routine procedure in fertility clinics for decades, and healthy babies have been born from multicellular embryos that have been stored at -196°C (-320°F) in liquid nitrogen for more than twenty years. In some cases, the mothers implanted with these embryos are close in chronological age to their babies. Frozen human embryos behave like Sleeping Beauty, which is a blessing for those who begin life in a flask of liquid nitrogen.

Freezing and thawing an entire adult is more problematic. This is the business of cryonics. Practitioners of this art try to freeze the cadaver, or just the head, as quickly as possible in the hope of turning the water within the tissues into a smooth glass. This vitrification process is designed to avoid the tissue damage caused by ice crystallization. The American Cryonics Society is headquartered in California – a strained pun lies there – and has contracts with a number of companies that store the corpses of humans and pets in liquid nitrogen. It is striking that nobody has agreed to have their head frozen before death, which means that the necessary technological advances will entail a full-blown Lazarus-style miracle.

An alternative natural method of suspended animation has evolved among tropical frogs that cope with the opposite climatic extreme from the frozen wood frogs when they respond to the summer dry season. These amphibians survive by shedding the outer layers of their skin to create a waterproof cocoon in which they rest in the soil for several months, until rainfall resumes. African lungfish extend this mechanism of mummification for up to four years by burrowing in the sediment beneath their shallow pools, encasing themselves in mucilage, lowering their blood pressure and pulse rate, and sitting tight as the mud bakes into a hardened clay.[9] We die when we are dehydrated. The tissues of a human who weighs 70 kg (155 lb) are soaked with around 40 l (85 pints) of water. Dehydration in the desert follows a standard pattern of a growing thirst, reduction in urine production, increase in body temperature and heart rate, drop in blood pressure, fainting, circling vultures, organ damage, kidney and liver failure, and death after losing around 10 per cent of one's body weight as the birds settle. An active body must be wet to keep blood thin enough to be mobile, to flush the kidneys and other organs, and to sustain the chemical ferment inside cells.

When physician-assisted suicide is not an option, or one decides against it, 'terminal dehydration' has been described as a relatively painless way to die for people capable of extraordinary resolve.[10] Before a related religious practice was outlawed in the nineteenth century, Japanese mystics from the Shungen-dô sect perfected a lengthy suicide process by eating an unappetizing diet of nuts, roots and resinous pine bark, secluding themselves for years at a time on a sacred mountain and, finally, being buried alive without food or water. A bamboo tube allowed the monk to breathe, and he rang a bell from his tomb from time to time to indicate his progress to those at the surface. The body of the ascetic, referred to as a *sokushinbutsu*, or 'embodied Buddha', would be exhumed after three years and placed on permanent display in a temple, mummified and dressed in clerical robes.[11] In an eighteenth-century short story by Ueda Akinari, an ancient *sokushinbutsu* is discovered in a village and is resurrected by rehydration. The villagers shower him with questions, but the priest is so absent-minded and shows so little sign of enlightenment that they abandon their Buddhist faith.[12]

Recognizing that the key to the extension of human life does not appear to lie in freezing or drying, space scientists have turned their attention to drug-induced hibernation as a plausible method of enabling cosmic safaris.[13] Even if it does not extend the lifespan of the astronauts, hibernation would obviate the horrific boredom of a long space flight and reduce the weight of provisions needed to keep a more active crew fed and watered. Humans do not hibernate, and it is difficult to find a good animal model whose lengthy snoozes might be appropriated for space travel. The first problem is that we are warm-blooded. Hibernation is much simpler for cold-blooded animals that have very limited control over their internal body temperature. Frogs just get cold. Small warm-blooded mammals that hibernate appear to deactivate their internal heaters in winter and stop fighting to keep warm. This is an energy-saving mechanism that requires the animals to shut off the mitochondrial furnaces within their fat tissues. Some bats come close to equilibrating with the cold air in their winter roosts, so that they are barely above freezing.

The dormouse is quite brilliant in its hibernation skills. It sleeps deeply for as long as eleven months in a year when the beech seeds that it favours are scarce, although it will awaken periodically to snack on food stored in its nests. During these times of exceptional ennui, it seems that the instance of apparent nonsense offered by the dormouse in *Alice's Adventures in Wonderland* was, in fact, close to a scientific description of the behaviour of the rodent: 'You might just as well say . . . that "I breathe when I sleep" is the same thing as "I sleep when I breathe"!'[14] Dormice, in common with other hibernating mammals, fatten themselves up in preparation for their deep slumber. This habit was exploited by Roman chefs, who confined the animals in special terracotta jars fitted with interior shelves to hold acorns and chestnuts, perforated for airflow and closed with lids to keep the dormice in the dark. With little to do but prepare for hibernation, the rodents plumped up, innocent of their preparation for post-mortem seasoning with honey and poppy seeds and roasting, as Petronius described in Trimalchio's immoderate feast in the *Satyricon* (AD 54–68).[15]

At the other end of the weight range of hibernating animals, bears – big and furry, rather than small and watery – shut down for a few months each year according to their circadian programming. Black bears and brown bears, or grizzlies, enter a state of dormancy for up to seven months. They do not eat, drink, urinate or defecate during this time, but maintain near normal body temperature. Undisturbed, their hearts beat very slowly inside the den, although they quicken in mothers when they give birth, before returning to a slow and erratic pulse as they nurse their cubs. Brown bears that hibernate have the same lifespan of twenty to thirty years as their close relative the polar bear, which does not. Both species can keep pacing around their enclosures for a decade longer in captivity, demonstrating that bears in zoos, like humans imprisoned for life, are not blessed by the natural release of death from the stress and cheerless monotony of their days.

A dwarf lemur from Madagascar is the only primate that hibernates. It lives in dry deciduous forests and sleeps in tree holes for up to seven months during the arid season, when fruit is very scarce. The body temperature of the lemur follows the air temperature during hibernation, fluctuating by as much as

25°C (77°F) in a single day.[16] The common feature of the various methods of hibernation, whether or not the body temperature of the sleeping animal falls, is a sharp reduction in metabolic rate. Captive dwarf lemurs live for thirty years, which is much longer than other small primates. Hibernation in other mammals is also associated with life extension, most notably in the little brown bat, which can live for 34 years in the wild, and the related big brown bat, whose lifespan exceeds forty years.[17] These American bats, as well as other bat species, live as much as ten times longer than mammals of comparable size that do not hibernate. Bats have unique metabolic and behavioural characteristics that favour a long life, but the preservative effect of hibernation is spread among other groups of mammals. In an experiment on Turkish hamsters, animals that went into hibernation after transfer to a cold room lived longer than those that were kept at a constant warm temperature and never hibernated. Some hibernators remain in a deep sleep or torpor for months, while others can be roused quite easily.

We are still talking about months of unconsciousness here, of course, rather than the thousands of years required for interstellar missions. Travelling at the speed of a space probe like NASA's *New Horizons*, we would enter interstellar space in 35 years, before globetrotting for 19,000 years per light year to reach the planets orbiting stars in our immediate galactic neighbourhood.[18] Space is just so very big and ablaze with cosmic rays that it seems very unlikely that there is a glorious future for its human exploration.

These facts of distance and danger do little to dull the enthusiasm of space scientists, however. In the absence of a natural mechanism of human hibernation, they have settled on 'synthetic torpor' as the best hope for keeping us fresh for extended voyages. But how might we encourage this? Experiments have shown that exposure to hydrogen sulphide – the rotten-egg gas – is effective at pushing mice into a state of suspended animation. This was seen as a breakthrough, at least briefly, because mice are non-hibernating animals like us. Gassing caused a drop in the metabolic rate of the animals, followed by a decrease in body temperature. In a less encouraging vein, this result was expected because hydrogen sulphide acts as a metabolic toxin. According to one of the experts in this field, 'The difference between a manipulation as you like to have it, on the one hand, and poisonous irreversible damage, is extremely narrow.'[19] On balance, then, hydrogen sulphide treatment does not seem any more likely to be embraced by astronauts than being banged on the head with a brick.

A more promising method takes us back to the space rockers of Hawkwind and to Lewis Carroll, whose mutual interests in psychotropic mushrooms are shared by neuroscientists. It appears that such mushrooms may send us into a state of extended calm that could let us reach the stars. Injections of minuscule quantities of a compound called muscimol into the brainstem of rats causes an immediate dilation of blood vessels and cooling of the brain, coupled with the display of electrical rhythms reflecting deep sleep.[20] Muscimol is the potent hallucinogen found in fly agaric mushrooms. Rats injected with it lose their sense of timing in maze experiments, as judged by their failure to negotiate a drawbridge that opens and closes at specific time intervals. Human 'psychonauts' often report alterations in the perception of time after consuming fly agarics, along with feelings of euphoria, out-of-body experiences and the distorted impressions of the size of objects known as dysmetropsia or Alice in Wonderland

Syndrome. This suggests that astronauts infused with muscimol could pass lengthy spaceflights in a state of mild intoxication and accompanying cerebral entertainments that might mitigate the extreme tedium of the mission. It is a reasonable bet, however, that they would emerge, like Mr Van Winkle and the Japanese monk in Akinari's tale, older and no wiser from their pods. Mushrooms might keep us occupied to the edge of the solar system, and then what?

The reason for my scepticism about life extension and space travel stems from my first job, in which I tended a graveyard in an Oxfordshire village, and a more recent appreciation that the Second Law of Thermodynamics explains absolutely everything. The graveyard responsibilities overlapped with my love of Hawkwind, but in a time before personal cassette players the only musical distractions during my labours came from the crows that nested in the horse chestnuts, and a tuneful robin that hunted for insects exposed by my mowing and raking. Hours of contemplation between the gravestones made it evident, in a way that my childhood had failed to convey, that everyone dissolves. A pair of gravediggers stopped by every week or so to dig perfect straight-sided trenches in the flinty soil. They dug with spades, following a rectangle marked with pegged string, and stood back afterwards, lit cigarettes and admired their handiwork. Samuel Beckett described the sorrowful path from obstetrics to gravedigging in *Waiting for Godot*: 'Astride of a grave and a difficult birth. Down in the hole, lingeringly, the grave digger puts on the forceps. We have time to grow old. The air is full of our cries.'[21]

The Second Law explains the need for gravedigging in scientific terms. It states, simply, that the universe and all its component bits are becoming more disordered as time passes.

Entropy, which is an expression of disorder, has increased every moment since the Big Bang, and eventually this process of heat transfer will leave the universe uniformly cold and dark. Long before this 'heat death' of the universe, all human substance will have dissolved and blended with the rest of the chemistry of the galaxy. This happens to us because we get older, our DNA becomes irreparable and the faulty proteins generated by these mangled instructions stop our cells from working properly. Then we die, which provides vultures and gravediggers with livelihoods.

Having dedicated my research career to experiments into the growth and reproductive lives of fungi – the sovereigns of decay – it is evident that I have indulged in more than the quotidian interest in death. Unfortunately, knowing a bit about what happens afterwards and revelling in the science of decomposition does not take much of the dread away from the inevitability of becoming food for mycelia. In his discourse on burial urns, the seventeenth-century writer Sir Thomas Browne noted, 'The long habit of living indisposeth us for dying,' which has led to pyramid-building and funerary rituals, is the certain root of monotheistic religions and has encouraged a relentless quest for methods of extending life.[22]

Alchemists sought the Philosopher's Stone, or elixir of life, which would turn base metals into gold and silver and endow the sorcerer with immortality. Turning to contemporary pseudoscience, the 'longevity escape velocity' is among the more ludicrous ideas, which rests on the delusion that medical advances could postpone death indefinitely if they increase life expectancy at a faster rate than the actual passage of time. The idea is that if modest technological enhancements every year can keep us just ahead of the scythe, 'sufficiently aggressive intervention' might keep us going for centuries.[23] Current approaches to life extension include calorie-restricted diets, vigorous colon cleansing, hormone therapies, vitamins, 'geroprotecting' drugs, probiotic yoghurts and natural medicines gleaned from plants, mushrooms and endangered animals. The injection of geriatrics with blood from youthful donors is another strategy, and a growing gaggle of 'inspirational speakers' promote mindfulness as a way to forestall the reaper. It is noteworthy that this preposterous circus has been entirely unsuccessful in drawing out the maximum human lifespan for one second.[24]

Vaccines and other marvels of Western medicine, along with increased agricultural productivity and modern sanitation, allow more humans to reach old age in our time, but the maximum attainable lifespan has not shifted very far since humans diverged from other species of great apes.[25] The British biologist Peter Medawar described ageing as 'an artifact of domestication' that was made possible 'by sheltering it [the animal] from the hazards of its ordinary existence'.[26] Historically, most humans died from infectious disease, starvation, physical injury and other hazards that we now sidestep in the developed world in favour of death from organ failure, cancer, dementia or more malleable expressions of comprehensive bodily exhaustion.[27]

Some of us will see ten decades; no one will face thirteen. Browne again: 'Generations pass while some trees stand, and old families last not three oaks.'[28] We wear out and, ready or not, with time flying into the future at 315 million seconds per decade, we will soon escape from all mortal concerns. Meditating on death in his *Natural History*, Pliny the Elder saw it as a blessed relief: 'nature has granted man no better gift than the shortness of life. The senses grow dull, the limbs are numb, sight, hearing, gait, even the teeth and alimentary organs die before we do.'[29] This 2,000-year-old expression of mindfulness deserves more attention as a counterpoint to the durable fear of death and our desperate attempts to deny its inevitability. To end with Beckett: 'They give birth astride a grave, the light gleams an instant, then it's night once more.'[30]

SEVEN

BOWHEADS

Centuries (10^9 Seconds)

The day began gloriously as she cruised north through the Bering Strait towards the Chukchi Sea. The sparkling water felt cooler and she could smell swarms of krill with every breath of air. She was a gorgeous animal, thirteen years old and already more than 11 m (36 ft) from blunt snout to streamlined tail. Other bowheads were calling with excitement about the richness of the food. Taking a deep breath through her blowholes, she clapped her nostrils shut, raised her flukes from the water and sounded. There was no immediate need for the dive; she was drawn on a whim to take a look around. Ten minutes later, in her eagerness for the surface, she had become distracted and thought the boat was further away. Then the metal spike sliced through the skin behind her left eye, slid into the blubber and split the muscle before striking the back of her skull. She heard the men shouting. She had never felt pain like this before. Ice-breaking could be hazardous in winter and left white scars behind her blowholes, but this was an agony beyond comprehension. Heart pounding, snout down and spine arched, she dropped into the vault below.[1] Swinging her massive head from side to side to throw off the nausea made for a disorderly descent. The year was 1888.

Some 8,000 km (5,000 mi.) away in Arles, another teenager, Jeanne Calment, was watching her uncle count the coins offered by a young man in his fabric shop. Neither man seemed happy with the transaction. Uncle Jacques asked for more money and the artist snapped two francs on the counter, 'C'est assez?' Tipping the brim of his straw hat to Jeanne and picking up the

Bowhead whale, *Balaena mysticetus,* whose lifespan can exceed two hundred years.

canvases, he said, 'Bonne journée, ma petite fille étrange.' What did he find strange about her, she wondered? He had a sickening smell, she thought, 'comme un cheval mort un jour d'été' – like a dead horse on a summer's day. The artist painted two canvases of sunflowers that afternoon and drank himself senseless in the evening. His friend Paul Gauguin helped him into bed.[2]

The whale lived for 92 years after the attack and was killed by native whalers off St Lawrence Island, Alaska, in 1980. Jeanne hung on for a while after that, dying in 1997 at the age of 122. No human has ever lived longer. The age estimate for the whale was based on the discovery of the tip of the nineteenth-century bomb lance embedded in her bone.[3] This missile was fired from a heavy bronze gun that was held and aimed like a rifle, and was designed to explode a few seconds after the sharp tip dug into the animal. In this case, it failed to detonate properly, and the long shaft with attached 'fishing line' pulled free of the blubber as the whale sounded.

Confidence in the age of Jeanne Calment has been shaken by the investigations of the Russian mathematician Nikolay Zak.[4] Without the gift of an ancient harpoon stuck in her skeleton, the claim that Calment was born in 1875 is based on census records and family photographs. Zak argued against the statistical likelihood of one woman living three years longer than any other human in history, showed several inconsistences in her account of her life, and supplied a very plausible rationale for fraud. Jeanne Calment was the mother of Yvonne, who was born in 1898. Records show that Yvonne contracted pleurisy and died in 1934 – or did she? Zak suggests that Jeanne died in 1934, whereupon Yvonne assumed the identity of her mother, so that 'Jeanne' was in fact just a year shy of a century when she died in 1997. The fact that the family would have avoided payment of inheritance taxes on properties owned by Jeanne by claiming that Yvonne had died provides the necessary incentive for the identity switch and subsequent cover-up. French genealogists, and the mayor of Arles at the time of Jeanne's (or Yvonne's) death, disputed these findings, but it is clear that they had their own biases in favour of the original record. If Zak is correct, the oldest well-documented human in history was Sarah Knauss from Pennsylvania, who died in 1999 after her 119th birthday.

'The days of our years are threescore years and ten,' or more, and sometimes a lot more, but we cannot creep beyond six score before we 'fly away' (Psalms 90:10). Bowheads can do much better. The oldest ones cruising the Alaskan waters at this moment were born decades before their holocaust, which began in 1848 and wiped out 90 per cent of the species in the region. Evidence for such longevity comes from the analysis of eye lenses collected from whales 'taken' by Alaskan Inuit during their annual subsistence harvests. The dating method relies on the natural conversion of amino acids, the building blocks of proteins, from one chemical configuration into another. Cells make proteins by joining amino acids in a precise sequence and spitting them out like strings of sausages on a conveyor belt. Amino acids come in two flavours, called the L-form and the D-form. These are shaped as mirror images, like our left and right hands, but only the L-form is used for the manufacture of proteins, including crystallins that form the glassy structure of lenses. As time passes, the L-form amino acids buried in the lens flip to the D-form. If we know the rate at which this happens, we can determine the age of the animal from the ratio of the two forms of amino acid in its lenses. Measurements of the proportion of the D-form of an amino acid called aspartic acid in a lens from a 15-metre-long (50 ft) male bowhead taken by villagers from Wainwright, Alaska, in 1995 suggested that it was 211 years old.[5]

When investigators provide an age estimate for an old animal, it is important to recognize that it is associated with a particular level of confidence. In the case of the bowhead from Alaska, the accuracy of the age estimate of 211 years is limited by the amino-acid dating method, and so it comes with a standard error, or statistical plus/minus, of 35 years. The whale might have been somewhat younger or, with equal likelihood, it might have been older. The best age estimate for this marvellous submariner suggests that it was born in the year that followed the end of the American War of Independence. I wonder how many more years it would have lived if it had escaped the Inuit?

The bowhead brain is very small compared with the enormous computers employed by toothed whales, and its hippocampus is particularly tiny.[6] The hippocampus takes care of long-term memory, which encourages the soothing thought that the ancient leviathan had forgotten the trauma of the whaling fleets from its early adulthood, rather than thinking, 'Oh shit, the bastards are back again' when it heard the approaching boat. There is some hubris in my attitude to the subsistence hunting of whales. Native people from the Arctic have lived this way for millennia and modernity has ruined their rich cultures in many ways, not least by melting the ice. The essential problem lies in the panoptic tragedy of human existence, as I have written elsewhere. Many of us should not throw stones at whalers because we live, most definitely, in glass houses, and we have consumed the flesh of sea creatures as elderly as the bowhead, with fresh tarragon, perhaps, and chilled white wine. A white fillet of orange roughy, hot from the grill, crusted with its own oils, may have lived in the depths of the Pacific Ocean for 150 years before its transfer to the refrigerated hold of a trawler. Not as smart as the whale, but brimming with its own sensitivity, this ancient animal died desperate to escape the net.

Age estimates for orange roughy and other deep-water fish are based on the study of their ear bones, or otoliths, which sit behind the brain and are components of the inner ear. Gravitational drag and sound waves displace the otoliths, whose motion is detected by sensory cells and provide the animal with its senses of balance and hearing. They look like tiny seashells with raggedy edges, and range from the size of a pinhead to the diameter of a wristwatch. There are two ways that they are used to determine the age of fish. First, they have annual growth zones that can be counted like tree rings, and second, they trap radioactive elements whose decay serves as a very reliable timepiece. For age analysis, the otoliths are cut or polished to reveal their interior growth zones, and embedded in resin on microscope slides. Ring counts show that orange roughy can live for more than a century, but it is difficult to resolve the complex layering patterns in otoliths from even older fish.[7] This is where radiometric dating comes in and we measure the ratio of lead to radium in the otolith core. Radium-226 (the number refers to its atomic mass) has a half-life of 1,600 years and decays to lead-210. Lead-210 has a half-life of a little over 22 years and decays through a series of short-lived elements to lead-206, which is a stable, non-radioactive isotope, or version, of the element lead.[8] The activity of lead relative to radium is a proxy for age – more lead as time flies – and this method showed that an elderly orange roughy caught off the coast of Tasmania in the 1980s was 149 years old.[9]

Radiometric dating shows that other deep-sea fish live as long as the bowhead whale. The record-holder is a pink-skinned rough-eye rockfish, whose otolith chemistry suggested that it was 205 years old – born off the coast of California at the close of the eighteenth century. There are more than one hundred species of rockfish in the family Sebastidae. Some can grow to 1 m (3 ft) in length and weigh as much as a dachshund. Unlike orange roughy, which are caught by drag nets that scoop everything from the sea floor, rockfish are caught on hooks by sports fishers. This seems more sustainable, but the overfishing of some species that is evident from the abundance of online 'rockfish recipes' has placed them in jeopardy. Other fish that keep swimming into a second century include grenadiers or rattails, sablefish, white sturgeon, the 'warty oreo' that may live for 130 years, and captive koi that bathe in their lovely Japanese water gardens for two hundred years.

Bony fish that reach the two-century mark appear to command the swift attention of the reaper, but at least one cartilaginous fish, the Greenland shark, adds another century before its expiry date. The oldest shark on record was 5 m (16½ ft) long and had an estimated age of 392.[10] Its Latin name, *Somniosus microcephalus*, refers to the sleepy behaviour of this animal (the Latin *somnio* means dream) and to its small head. The skull houses a tiny brain that matches the weight of a single AAA battery or the brain of a rabbit, which is alarmingly small for an animal as big as a horse. Horse brains are fifty times bigger.[11] All sharks have small brains in relation to their body size, but the PC that runs the Greenland species is strikingly petite. The brain of a great white shark of the same body weight is twice as big.

The explanation for this anatomical idiosyncrasy lies in the relative simplicity of the animal's lifestyle. *Somniosus* passes the decades by swimming very slowly in deep, frigid water, feeding on fish and the occasional seal if it chances upon one sleeping in the water. Annual catches of tens of thousands of sharks supported a market for its liver oil, which was used as an industrial lubricant until the 1960s. Icelanders continue to catch the whale for its meat, which is consumed as a delicacy called *kæster hákarl*. Even after months of preparatory fermentation and drying, *hákarl* has a strong smell of ammonia that non-natives find repellent. More sharks are killed today as by-catch of halibut fishing, and it is listed as 'near threatened' by the International Union for Conservation of Nature pending better data on population sizes.

The remarkable longevity of the Greenland shark has been established by looking at the activity of radioactive carbon in its lenses, rather than the ratio of the different forms of amino acid that unmasked the ancient bowheads. Radiometric dating of eye lenses has become a very popular method for determining the age of animals. This method makes elegant use of the elevated levels of carbon-14 in the tissues of every living thing, produced by the above-ground tests of nuclear weapons from the mid-1950s to the early 1960s.

Carbon-14 is produced when nitrogen-14 atoms combine with neutrons. Neutrons are the uncharged subatomic particles that accompany protons in the nucleus of an atom. They are discharged as free particles when cosmic rays collide with atoms in the atmosphere, and this natural process fuels the creation of carbon-14. Carbon-14 reacts with oxygen to form radioactive carbon dioxide, and this enters food webs when it is absorbed by plants, algae and bacteria that carry out photosynthesis. Carbon-14 is also produced in nuclear reactors and in the airburst following the detonation of a nuclear bomb, which is how the weapons tests doubled the concentration of carbon-14 in the atmosphere. Carbon-14 is unstable, and its decay produces nitrogen-14 again; it has a half-life of 5,730 years. If carbon-14 is trapped in the tissue of an organism, meaning that it cannot get out and new carbon-14 cannot get in, its activity provides a measure of how long it has rested there.

The natural level of carbon-14 in the atmosphere is relatively constant, which provides a baseline measurement, and the bomb pulse serves as a timestamp for organisms born before and after the early 1960s. Levels of carbon-14 began to wane after 1963, which means that anyone born in the mid-1960s, or later, is less radioactive than those of us alive during the festivities organized by the original nuclear powers. The choice of tissue for ageing studies affects the interpretation of the bomb pulse, or its absence. Adult human tooth enamel provides a hallmark for animals born in these singularly (we hope) radioactive years because it is not replaced after it has formed, and tendons show similar patterns of radioactivity. Teeth and tendons are useful in forensic investigations to determine the age and date of death of an individual. The same radiometric methods are also useful for validating vintage wines and works of art.[12] Grapes and oil paints used by fraudsters cannot conceal their youth.

Any long-term health effects of the bomb pulse are unclear. Carbon-14 is not as dangerous as other radioactive elements produced during the weapons tests, and the quantities in our tissues are very low. Some studies have estimated that more than 2 million people could die from cancer caused by the fallout, but it is impossible to discriminate between those casualties and the far greater number of cancer cases arising from other causes. In a counterintuitive and controversial vein, it has been suggested that low doses of radiation may actually extend lifespans by reducing deaths from cancer. Advocates of this 'radiation hormesis' idea point to particular readings of demographic data from people exposed to radioactivity through industrial accidents and the bombings of Hiroshima and Nagasaki in 1945. They propose that by stimulating mechanisms that repair cells, low doses of radiation may favour the body with continuing protection. Again, however, cancer is such a prevalent cause of mortality, and comes in so many flavours, that it is very difficult to measure the impact of low doses of radiation on the disease.

Greenland sharks do not breathe air, and live a long way from the places where the weapons were tested. Radioactive fallout reached their ocean depths nonetheless, via algae and bacteria that absorbed the 'hot' carbon dioxide after it dissolved in the sea, larger planktonic organisms that ate the algae, fish that gulped the plankton, and so on. Nothing in biology was immune from the fallout. As early as 1955, scientists in New Zealand measured rising levels of carbon-14 in the atmosphere, in seawater and in trees. The increase was quite modest at that time, amounting to around 5 per cent above the historical baseline, and the scientists sounded hopeful in their published report: 'Should atomic weapons testing cease, the C^{14} specific activity of the atmosphere would begin to return to the pre-atomic bomb level.'[13] This did not happen, of course, and so New Zealanders were contaminated along with the giant fish going about their business in the Arctic Ocean.

Remaining in cold seawater, not far from the home of the Greenland shark, we find the oldest individual animal of all, the ocean quahog, *Arctica islandica*, a large edible clam. (Colonial animals of even greater age will be featured in Chapter Eight.) One specimen collected from the waters around an Icelandic island in 2006 was 507 years old. Born in the middle of the Ming Dynasty, it was named Ming the Mollusc by a journalist writing for the *Sunday Times*, and also, rather vacuously, Hafrún – 'the mystery of the ocean' – by the Icelandic researchers who collected it.[14] I disparage the Icelandic name because there is no mystery attending any clam, however elderly; it just took scientists a long time to catch up with this one, kill it and count its rings. And counting rings, the narrow growth bands in the shell, was the method used, alas, to date poor Ming. Analysis of the width of successive growth bands in even older shells, left by animals that died long before they were collected, provides an archive of water temperature extending back as far as the seventh century. The growth rate of these ancient clams changed during the late thirteenth and early fourteenth centuries, corresponding to alternating periods of warming and cooling that characterized the transition from the Medieval Climate Anomaly, or Warm Period, to the Little Ice Age that followed.

This overview of antique animals suggests that there is something very beneficial about living in the sea and, particularly, doing so in cold, deep water. This is borne out by comparing the longevity of rockfish species that live at different depths in the Pacific Ocean.[15] There is a clear correlation between depth and longevity. Rougheye rockfish, with the 205-year lifespan that was mentioned earlier, live between 150 and 450 m (500–1,500 ft) below the surface. The deep-sea lifestyle comes with high water pressure, low light, reduced oxygen and scarce food. Temperature drops with depth, too, and Ming the Clam, who lived at 80 m (260 ft) on the North Icelandic Shelf, demonstrated that cold water may be more important than the other variables associated with depth. More important still than the cold marine lifestyle itself is the slow and steady metabolism that is essential to thrive in these conditions. This physiological attribute is shared with the oldest land animals.

The oldest of the sun-baked residents of the biosphere are reptiles, and the tuatara and tortoises deserve special commendation. We encountered the tuatara, with the pronounced third eye sitting on top of its head, in Chapter Four in our discussion of circadian rhythms. Tuataras live on some of the smaller islands of New Zealand, and their name, from the Maori language, refers to their spiny backs. The tuatara is the only surviving species in a lineage of lizard-like animals called the Rhynchocephalia that evolved in the Triassic Period. Several features of tuatara anatomy distinguish them from lizards, including the presence of abdominal ribs and the absence of external ears. Individuals live for more than a century in captivity, and one gentleman in Invercargill, called Henry, became a father at the age of III, with his eighty-year-old partner Mildred. Some herpetologists believe this species can live for two hundred years.

Giant tortoises can certainly prosper long into their second centuries. Galapagos tortoises probably live as long as the tuatara. Populations of these animals were decimated in the eighteenth and nineteenth centuries by the combined attentions of sailors from military vessels and whaling ships that stopped off to replenish their supplies of food and water. The tortoises were gathered by the mariners and stacked upside down and alive in the holds of ships as a source of fresh meat as they proceeded on their voyages. Miners on their way to California during the Gold Rush in the 1840s added to the exploitation of these reptiles, collecting them on their way north after rounding Cape Horn, with more than 5,000 km (3,000 mi.) to go before they reached San Francisco.[16] Permanent settlers on the islands used them as a convenient protein, too. Charles Darwin was entranced by the different subspecies of tortoise when he visited the Galapagos in 1835. A tortoise called Harriet, who died in a Queensland zoo in 2006, was reputed to have been collected by him and taken to Brisbane by an officer who left HMS *Beagle*. This story was discredited when a stuffed specimen of a young Galapagos tortoise discovered in the British Museum was identified as Darwin's pet. It seems he had taken it home, whereupon it perished in the cold climate.

Moving to Seychelles in the Indian Ocean, we encounter the Aldabra giant tortoise, which has the longest verifiable lifespan of any land animal. A nobleman called Jonathan hatched in 1832, or perhaps a little earlier, which makes him at least 188 at the time of writing. He was taken from Aldabra Atoll to live as a guest of the government on the island of St Helena, in the South Atlantic, when he was fifty. This casts him as a reptilian echo of Napoleon Bonaparte, who was exiled to the same volcanic island at the age of 46. Jonathan fared the better of the two. Other tortoise species have made it past the century mark, including the Greek tortoise, which lives to at least 127 in captivity.

Slow and steady wins the race, as Aesop said, and being an ectotherm (whose temperature rises and falls with the environment) and having a slow heart rate seems to be part of the key to life extension.[17] The Galapagos tortoise maintains a low metabolic rate and has a resting pulse of 6 to 10 beats per minute. Over two hundred years this equates to somewhere between 600 million and 1 billion heart contractions, which aligns with the magical number of heartbeats accorded to most animals.[18] The apparent value of the slow burn does not auger well for life extension in humans. For animals with such an active metabolism, we already live a few decades longer than we might predict by inspecting the rest of zoology. According to the loose 1 billion heartbeats guide, an individual human should only live about thirty years. This means humans are remarkably long-lived, and very unlikely to live much more than we already do. Analysis of the genomes of bowhead whales and similarly durable species has uncovered modifications to genes that control cell division and DNA repair. If these turn out to enhance the suppression of cancer, they may be part of the reason why some animals live so long. Yet there seems little value, from an evolutionary point of view, in any animal living for hundreds of years. Bowheads reach sexual maturity in their twenties and have ample opportunities to bestow their genes upon calves before they pass the century milepost.

Every species is engineered for the swiftest transmission of their genes allowed by their physiology and environment. There is no advantage in waiting. Each animal has a characteristic survivorship curve that shows the number or proportion of individuals surviving at each age. Humans and other large mammals that have few offspring do quite well in their early years, survive early adulthood and enjoy a few years of sexual maturity before they begin to drop like flies.[19] These species contrast with marine invertebrates, such as oysters, which produce huge numbers of offspring that are mass-slaughtered as larvae. The minority of survivors tend to live for quite a long time. Whatever reproductive strategy evolves, only a tiny proportion of individuals make it to the oldest ages. Such was the fate of Jeanne Calment (arguably) and Jonathan the tortoise (definitely), who outlived all their contemporaries by decades. Some animals live for more than 3 billion seconds, but as the arrow of time proceeds, nothing with a brain abides for 30 billion.

EIGHT

BRISTLECONES
Millennia (10^{10} Seconds)

Most of the age intervals delineated in these chapters are our inventions. Days are the sole exception, in the sense that other organisms recognize Earth's spin every 86,000 seconds, and behave according to their inbuilt circadian clocks. Years are real, too, as the pulse of the planet's solar circumnavigation, but their 30-million-second cycles are registered indirectly from cumulative changes in day-length and the weather. Earth days and years would be measured by observant aliens, whatever calendars they follow at home, just as we calculate the solar days and years of the other planets. Everything else is completely made up, even the second, of course, to which this book beats over so many orders of magnitude. As Thomas Mann wrote in *The Magic Mountain*: 'Real time knows no turning points, there are no thunderstorms or trumpet fanfares at the start of a new month or year, and even when a new century commences only we human beings fire cannon and ring bells.'[1]

This make-believe quality of time is at odds with our experience of the arrow, which we follow into the future, watching the Sun rise and set, looking at clocks, consulting the calendars on our phones and planning to buy bread before the shop closes. Life requires us to press on with things in this manner, paying attention to time as if weeks, months, years and so on *do* exist. Whether we race through life without an opportunity to pause, or slow the pace through meditation, everyone scuttles across a bridge too short between the infinite past and a future of nothingness. Millennia are hefty chunks of time compared with this fleeting affair. The 30 billion seconds that tick by in ten centuries are enough to make a mockery of every pretence of permanence. There is an opportunity for poets and dictators to make a lasting post-mortem impression, but is anyone remembered for anything more than their role in a dubious legend after a thousand years?

The good news about millennia from a zoological standpoint is that they erase suffering. Memories of personal or familial terror can persist for a century in whales and men, but not for much longer. We memorialize the casualties of wars, famines, plagues, earthquakes and other catastrophes, but sooner or later none of these tributes attracts any sympathy. Life is too short to feel mortified by the disasters of history. We have more urgent challenges. Unconscious lives have a lot more at stake on this extended timescale, and in this chapter we return to plants, consider the longevity of fungi that form mushrooms, and explore the possibility of immortality in jellyfish and other soft-bodied animals.

Bristlecone pines are the oldest-lived individual organisms. (We will wrestle with the nature of individuals shortly.) One of them in eastern Nevada, named Prometheus, was at least 4,900 years old when it was cut down by an enthusiastic forestry student in 1964. The student knew he was felling an ancient tree, because he had already counted more than 3,000 annual rings in core samples taken from other bristlecone pines in the same stand. He wrote, 'To facilitate compilation of a long-term tree-ring chronology . . . one of the larger living bristlecone pines was sectioned.'[2] The U.S. Forest Service had granted permission, and may have regretted doing so once the student reported that he counted 4,844 rings with an average width of 0.5 mm. It certainly lamented the ensuing publicity. With allowance for missing rings caused by years of extraordinary drought, it was estimated that this pine had germinated in the thirtieth century BC, around the time the first wooden version of Stonehenge was built. A tree named Methuselah grows in the White Mountains of California and is a few decades shy of this record today. Its exact location is kept a secret.

The Sierra Nevada boasts 3,000-year-old giant sequoias (redwoods) and junipers; bald cypresses in North Carolina and figs in Sri Lanka live for more than 2,000 years; and two dozen tree species crack the 1,000-year mark. The majority of these trees are species of conifer that bear cones. The oldest flowering plant is the African baobab, with a maximum lifespan that exceeds 2,000 years. An exceptionally large example in Zimbabwe reached the age of 2,450 years before it toppled in 2011. Baobabs lack the clear annual rings of conifers, and have been dated from the levels of radioactive carbon-14 in their wood. Climate change is taking its toll on these monsters, and many of the oldest and largest trees across their geographical range have collapsed in recent years. Bristlecone pines are also threatened by warming, primarily through competition from other trees that are able to colonize higher elevations as the treeline edges towards the mountain peaks.

Ancient trees are climate beacons because they are stuck in the same place for millennia, exposed to each shift in the local environmental conditions, and grow every year in perfect step with temperature and rainfall. Bristlecone pines and other species that form conspicuous annual rings provide a detailed archive of these short- and long-term changes in weather and climate. Anthropogenic (human-caused) climate change is written in the widening of annual rings in the bristlecones since the middle of the last century.[3] By aligning sequences of tree rings from a combination of living and dead bristlecone pines, investigators have been successful in assembling 9,000-year records of tree growth, and they hope to extend this to create an 11,200-year archive of ring patterns. Trees growing that long ago would have experienced the rebound in global temperatures following the period of rapid climate cooling called the Younger Dryas.[4]

Bristlecone pines and other long-lived plants last ten times longer than the oldest animals: 160 billion seconds for the pine, versus 16 billion seconds for the Icelandic clam of the previous chapter. How do trees do this? Their continuous or indeterminate growth pattern is crucial. This provides plants with a high level of developmental responsiveness to the environment that can compensate for their immobility. By growing continuously when the environmental conditions are permissive, trees spread their roots to tap more water from the ground, branch into the air to display more and more leaves to the sky, grow new branches in response to limb loss and develop fresh sex organs in cones and flowers throughout their lives. Even though a bristlecone pine does not reproduce every year, its lasting fertility allows it to release huge numbers of seeds across the millennia. It seems likely that natural selection has rewarded this fecundity by configuring internal life-sustaining processes that are unnecessary for plants with shorter reproductive lives. Such protection includes a suite of biochemical defences to rebuff pests, natural 'weedkillers' to prevent competition from the seedlings of other plants, and hormonal controls to coordinate the activities of such a massive organism.[5] Some combination of these botanical blessings keeps the vegetables going for that extra order of magnitude of seconds.

Being immobile is an asset for a plant growing in a supportive and stable environment. Even though the ground is frozen for many months in the Great Basin where the bristlecones grow, and baobabs experience extreme drought across Africa, the repetition of these circannual conditions is a key to survival. The physiology of the trees is adapted for periodic freezing or baking, and year after year they deepen their relationships with the communities of microbes in the soil. They become the lynchpins of their immediate environment. Trees are environmental managers, stabilizing and enriching the soil, building on their success as time passes. Through this intimacy with their physical surroundings and other organisms, these giants shrugged off the swings and roundabouts of climate variation to which life was accustomed in recent millennia. And then it came . . . the climate catastrophe of this century, temperatures rising slowly, tenth by tenth of a degree, to confound even the champions of botany. Because everything has its limit.

Some plant seeds can remain viable for more than a millennium if they are preserved under very dry conditions. Rehydrated lotus seeds germinated after being buried in a lake bed in China for 1,300 years, and 2,000-year-old Judaean date palm seeds rejuvenated after they were discovered by archaeologists excavating the Masada fortress in Israel. Russian scientists have demonstrated that tissues of a plant called the narrow-leafed campion can extend the viability record by another order of magnitude.

They found that the embryos dissected from immature fruits of this species found frozen in Siberian permafrost grew on sterile medium in test tubes. After transplantation into potting soil, these seedlings grew into healthy adult plants that set their own seeds. Carbon dating revealed that the original fruits were 30,000 years old. In a frozen state, the DNA from this flower from the Arctic tundra had held up for 1 trillion seconds.[6]

Bristlecone pines and baobabs grow as individuals in the sense that we know where the plant begins and ends. All the cells in all the root tips of one pine contain the same genome, which marks them as belonging to the same plant. Clones that live as separate organisms complicate matters, although we continue to regard tomato plants produced from cuttings as individuals. Physical separation is enough to settle the question in this case. Things become more convoluted when we look at a colonial plant with several shoots sprouting from a single interconnected network of roots. Quaking aspen colonies in the western United States can spread over many hectares in this way, and the famous giant Pando colony of trees in Utah covers more than 40 ha (100 acres).[7] Stems in the colony have an average age of 65, but the root system is far older. Quaking aspens come in male and female versions, with male catkins shedding pollen and females producing cottony seeds. Pando is an all-male clone, bearing the same genome across its entire territory, which is important for estimating its age. If the colony were intermixed with female trees, or with other males, it would be difficult to disentangle the history of one colony from another. The purity of Pando suggests that it may have been established at the end of the last glaciation, which means that it might have been spreading for 80,000 years.

Neptune grass is a kind of seagrass that grows in shallow water in the Mediterranean. The roots of this plant expand slowly over the sandy sea floor for thousands of years, producing lush marine meadows. Genetic analysis reveals that a pair of meadows separated by 15 km (9 mi.) in the waters around the Spanish island of Formentera belong to the same clone. To cover this range, the seagrass must have been growing for between 80,000 and 200,000 years, or even longer if the colonies sprouted from an older plant that was split when the island was formed by falling sea levels.

Rather unassuming shrubs can also reach into multi-millennial lifespans, including King's Lomatia, a native of Tasmania, and a creosote bush in the Mojave Desert called King Clone.[8] Although the Tasmanian plant grows as individual shrubs with separate root systems, every wild plant of this species (*Lomatia tasmanica*) belongs to the same clone. Individuals grow for three hundred years, but fresh roots and a stem can form when a branch is split from the parent plant. Radiocarbon dating provides an age estimate of 43,000 years for fossilized wood belonging to the clone. King's Lomatia evolved as a hybrid of two different species of evergreen plant, and found itself with mismatched chromosomes that rendered it unable to reproduce sexually. Because of this dynastic dead end, King's Lomatia is listed as critically endangered. The Mojave creosote bush, meanwhile, has expanded in a ring around a bare centre. A combination of growth-rate measurements and the carbon dating of dead wood taken from the colony suggests that it may be 11,700 years old.

Fungi grow like creosote bushes as their microscopic filaments fan out from a central spot to form a circular colony of ringworm on someone's scalp or a fairy ring of mushrooms in cow pasture. Most mushrooms require a mating reaction between a pair of colonies, or mycelia. This muddies the idea of individuality a little, because one of the partners may be younger than the other, but the radius of the colony does provide a rough measure of the age of the oldest fungus in the marriage.

Colonies of giant puffballs form particularly impressive fairy rings in the remaining scraps of native grassland in the American West. According to their traditional beliefs, the Niitsitapi (Blackfoot) people imagined that these white globes, or *kakató'si*, were created by fallen stars.[9] They burned them as a ghost-repellent, and painted the fruit bodies as white circles along the bottom edge of tipi covers to symbolize the birth of life from the dark earth. There does not seem to be an inherent biological age limit for these colonies. As long as the filaments get a sprinkling of rain every so often, they continue to forge through the prairie soil a few centimetres a year. Early in the twentieth century, puffball colonies with a diameter of 200 m (656 ft) were measured on undisturbed areas of shortgrass prairie in eastern Colorado. These must have been growing for centuries, long before the *Mayflower* dropped anchor off Cape Cod. Back then, the most significant threat to these mycelia came from the hooves of migrating buffalo. Other mycelia are of greater age, notably the famous honey mushroom of Oregon that covers 10 sq. km (nearly 4 sq. miles) and was growing in the fourth century BC, when Plato composed his *Symposium*.

Fungi that partner with algae in lichens are older still. As a student, I spent a wet week in Wales measuring map lichens on boulders in Snowdonia, with the aim of discerning growth trends related to light exposure. The map lichen is a common species that grows on exposed rocks in cold locations. It is a pretty organism, with a yellow background that is fractured into small, black-outlined patches, lending it the look of a map of counties. It is described as a crustose lichen, which means that it is very flat and is stuck hard to the rock surface. (Other kinds of lichen hang from tree branches and are anchored to their perches by little discs resembling octopus suckers.) One of the characteristics of the map lichens we found in Wales was their increasing size towards the tops of the more angular boulders. The popularity of these sites as perches for birds of prey and species of crow suggested that the lichens were being fertilized on the biggest rocks. This led us to begin taking moisture samples from the boulders after it rained, to measure nitrogen levels. There seemed to be a limit to this growth stimulus, because the tops of some rocks were bare, with the lichens growing off-peak like the tonsures of monks. This was interesting, given the usefulness of lichens as indicators of airborne nitrogen pollution from fossil-fuel combustion, industrial activity and intensive farming. It seemed the birds were providing too much of a good thing.

These lichens were much more noteworthy than I recognized back then, as we slid over the rocks, hand lenses strung around our necks and precision dividers and notebooks at the ready. Published measurements of extremely slow growth rates, falling well below $1/10$ mm a year, suggest that individual lichens began life on Arctic rocks many thousands of years ago. The oldest specimens are documented from Colorado, Swedish Lapland, Baffin Island and . . . drum roll . . . the Brooks Range of northern Alaska, where a lichen has been accorded an age estimate of 10,000–11,500 years.[10] Map lichens can grow for twice as long as bristlecone pines, but because they expand like mushroom mycelia, we tend to think of lichens as colonies rather than individuals, so the comparison with the pines may be unfair. Younger lichens on headstones are often as old as the date of departure chiselled into the stone.[11] During our annual summer holidays in Lincolnshire in the early 1970s, my grandmother took us on outings to her favourite graveyards. Trudging across the boggy soil, reading the lichen-splattered stones, we marvelled at the early demise of so many Victorian children. 'Rest in Peace – Our Dear Little Millicent – Struck Down by Agonizing Boils', and so on. One reason we enjoyed these entertainments came from the feeling of liberation elicited by the fluke of being above ground, while the less fortunate rotted in the damp soil beneath our feet. But *Schadenfreude* in a cemetery is fleeting; graveyard visitors become residents and lichens perch on their headstones.

Glass sponges that grow at great depths in very cold water outlive lichens. These animals have silica skeletons that function as elaborate scaffolds for their squishy tissues. A bizarre species that grows at depths of up to 2 km (1¼ mi.) constructs a giant spine of silica that extends into the seabed to keep it in one place. The sponge tissues grow around this stalk, and older animals look a bit like bulrushes on the sea floor. Layer upon layer of silica is deposited on the surface of the spine as it elongates and increases in girth. Chemical changes in these ultra-thin layers occur in response to variations in the temperature of the seawater, and the spines can be dated by matching those patterns to climate records established by geologists. Analysis of the spine of one sponge dredged from the abyss of the Okinawa Trough in the East China Sea, between Japan and Taiwan, gave an estimated age of 18,000 years.[12] More impressive than the age of this animal is the astonishing appearance of its glass spine once it was cleaned for analysis: it was 3 m (10 ft) long and 1 cm (½ in.) in diameter – longer than an Olympic javelin and as clear as a fibre-optic cable.

Whether this sponge is the oldest individual organism on record or an ancient colony of cells is mostly a matter of opinion. The colonial designation is supported by the observation that the soft parts of some types of sponge can survive the disruption of being rubbed through a sieve. The clumps of cells begin to reaggregate immediately and commence the process of regenerating the whole animal. On the other hand, sponges have much greater cellular specialization than we used to recognize, with distinctive sets of cells forming an external skin, driving water through the porous structure of the sponge, secreting the skeleton and crawling around to unclog the channels for water flow. This division of labour echoes the anatomy of more complex animals.

Corals appear to be less ambiguous examples of colonial animals, because they are made from thousands of polyps. Each polyp is structured like a miniature sea anemone, and is filled with onboard algae, or zooxanthellae, which sustain the colony by photosynthesis. The situation is not as straightforward as it seems, however, because the coral polyps are strung together at the base by tissue that allows them to share nutrients across the colony. Like deep-sea sponges, black corals live for millennia, and growth-rate measurements and carbon dating have indicated a maximum age of 4,265 years for a Hawaiian species.[13] So, the oldest animal is a glass sponge, a black coral or the Icelandic clam: you decide.

Scientific interest in the oldest plants, fungi and animals is concentrated in the possibility that we might learn something about living longer by figuring out how they do it. Using the great age of trees as an argument for conserving the habitats in which they live seems to be secondary to the possibilities of human life extension, although, in the end of course, they amount to the same thing. Nobody questions the importance of conserving the pyramids of Egypt and yet it is difficult to get people to care about the grizzled trees that were born at the same time in the Sierra Nevada.

The human obsession with longevity, or, rather, fear of death, means that each discovery of an odd slice of superannuated biology gains many points on the scale of newsworthiness. This explains why the so-called immortal jellyfish reaches well beyond the celebrity status of old trees and tortoises. Jellyfish undergo a series of metamorphoses beginning with the growth of a fertilized egg into a flattened planula larva. The planula swims in the water column before it attaches to a rock and transforms itself into a polyp. When the polyp is fully grown it releases tiny ephyra larvae with radiating arms. The ephyra larvae undergo the final remoulding into adult jellyfish, or medusas, which reproduce by shedding eggs and sperm cells. After reproducing, the adults of most kinds of jellyfish disintegrate, but the 'immortal' species (*Turritopsis dohrnii*) escapes death by reverting to its juvenile polyp form and resuming growth. An individual jellyfish cannot survive for long in any of its various forms, but this one prolongs the expression of its genome by reversing its life cycle. F. Scott Fitzgerald provided a human parallel in his short story 'The Curious Case of Benjamin Button', in which the eponymous hero is born an elderly man, ages backwards and ends his life in infancy, bereft of memory: 'Then it was all dark, and his white crib and the dim faces that moved above him, and the warm sweet aroma of the milk, faded out altogether from his mind.'[14]

Pond hydras, which are relatives of jellyfish, were celebrated as immortals during the height of their fame in the eighteenth

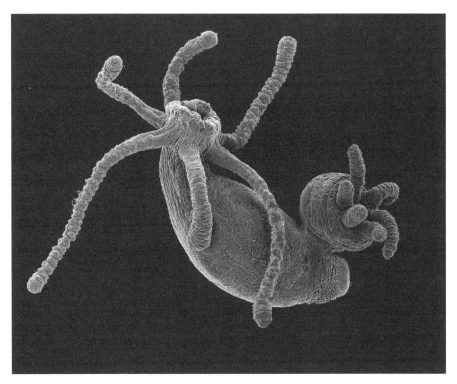

Hydra is related to the Portuguese man-of-war and is used in research on ageing. The animal in this electron micrograph image is giving birth to a clone from its flank.

century. Abraham Trembley, whose observations on the explosive stingers of these animals were mentioned in Chapter One, became famous for his work on the regenerative powers of hydras. He demonstrated their resilience to wealthy patrons in France and London by slicing the tiny animals in half, or lopping off their tentacles, and revealing that the separated tissues survived and were capable of growing into adults. Recent experiments have supported the general impression that hydras resist ageing.[15] Kept in culture dishes, free from the predators that hunt them in their outdoor ponds, hydras show no increase in mortality for at least four years. During this time they reproduce asexually by developing buds that detach from their parents when they are capable of feeding on their own. There is some evidence, though, of a long-term decline in bud formation, which may indicate that the animals are flagging. Four years may be a long time for a hydra, but it is not the stuff of legends.

Prospects for biological immortality improve if we pay attention to the continuity of genes rather than the survival of organisms. Taking this long-term view, we are all the couriers of information that will, someday, bookend the entire history of life, from the first to the last cell. Natural selection is blind to the persistence of the individual once it has fulfilled its genetic mission; Mother Nature does not give a hoot whether an organism drops dead or grows old after transmitting its DNA. The positive side of her indifference does not mean that any individual can hang on indefinitely, though. It might seem possible for a lichen to keep spreading over a boulder forever, but that rock will shatter someday or be submerged by the rising sea. The briefer exercises of geology place a maximum time limit on every local ecosystem, allowing a forest to sprout as an ice sheet recedes, for example, only to scrape it away with the next glaciation. The more prolonged formation and erosion of mountain ranges and separation of continents fill the epochs of geology, which consume tens of thousands of lifetimes of the oldest individual animals and hundreds of thousands of shorter occupancies like ours.

NINE

BASILOSAURS
Millions of Years (10^{13} Seconds)

Evolution permeates all the timescales in this book, from the generation of its raw material through split-second errors in DNA copying to the expansion of the tree of life over billions of years. Upgrades in the genetic make-up of a bacterial cell, or viral particle, spread like wildfire because these microbes reproduce so swiftly. When something useful arises among the mishmash of mutations, this serendipitous sliver of DNA is more likely to be conserved and carried into the future than less worthy survival instructions. This process steers the development of bacterial resistance to antibiotics and the birth of newfangled viruses unruffled by vaccines. Evolution proceeds via changes in the prevalence of different versions of genes in populations of microbes and macrobes. As genes are rejiggled, life rearranges itself along new branches, spreading like Jackson Pollock's drip paintings, as new species descend from their ancestors, allowing something like a hippo to be remodelled into something as monstrous as a whale.

Evolution crafted a whale from a relation of the hippopotamus, which probably looked more like the tiny mouse deer or chevrotain found in tropical forests today.[1] This was accomplished in about 8 million years, which seems very fast when we

Walking whale, *Ambulocetus natans*, who lived in the Eocene Epoch more than 40 million years ago.

recognize that it took this long to sculpt humans from an ancestral ape that probably looked a lot like us in the first place. The separate lines of descent that spat out chimpanzees and humans from a hairy forerunner over a few million years seem effortless compared with modifying a land animal into an ocean-going leviathan, but this is how it happened.

Hippos have four toes per foot, each capped by a hoof. They are even-toed ungulates, classified in the same group, called artiodactyls, as pigs, camels and cows. A horse stands on a single toe with a hoof, which makes it an odd-toed ungulate. Fossils of an extinct Indian artiodactyl, called *Indohyus*, offer a glimpse of the historical connection between land mammals and whales.[2] This animal was the size of a racoon and was built for wading in shallow water. Evidence for this amphibious lifestyle comes from its limb bones, which are thickened on the inside, leaving a narrowed central cavity for the marrow. In humans, this increase in bone density, or osteosclerosis, is associated with bone cancer and other disorders, but it kept *Indohyus* weighed down in the water. Hippos, which are described as bottom walkers, also have osteosclerosis. Chemical analysis of the tooth enamel of *Indohyus* confirms that this little animal spent a lot of time in freshwater habitats. It probably browsed on plants growing along riverbanks and submerged in the water, and foraged for shellfish and other invertebrates. This does not sound like a whale, but its fossilized teeth, skull and middle ears are shaped like those of a cetacean.

Indohyus lived around 48 million years ago, in the middle of the Eocene Epoch, when the tectonic plate carrying India crashed into the Eurasian Plate and thrust the Himalayas skyward. Tectonic plates, which are made from slabs of the Earth's crust and underlying mantle, move around the planet at a speed of a few centimetres per year. Collisions between plates cause mountains to rise a few millimetres per year, which amounts to half a metre (1½ ft) in a human lifetime, and kilometres every million years. The contrast between the achingly slow processes of geology and their spectacular results allowed the geologist Charles Lyell to argue in the 1830s that the Earth was immensely older than the span of a few millennia recommended by the Bible. And this revelation inspired Charles Darwin to begin to imagine how life might have diversified through the accumulation of incremental changes in form and function: 'He who can read Sir Charles Lyell's grand work on the Principles of Geology, which the future historian will recognise as having produced a revolution in natural science, yet does not admit how incomprehensibly vast have been the past periods of time, may at once close this volume.'[3]

In the epochs opened by Lyell, Darwin imagined that whales might have evolved from bears that became 'more and more aquatic in their structure and habits'.[4] The link between bears and whales was too fanciful for his critics, however, and Darwin struck it from the second edition of *On the Origin of Species*. He was right, of course, in his conjecture about the transformation of land mammals into whales. If only he had guessed 'deer' rather than bear. During Darwin's voyage on HMS *Beagle* the first fossils of a whale, which had been discovered in Louisiana, were described at a meeting of the American Philosophical Society in Philadelphia. Vertebrae the size of 9-litre (2-gallon) buckets were strung out in a line, encouraging the reference to a 'sea monster' in the original description. Because it was thought the beast might be a marine reptile like a plesiosaur, the extinct animal was named *Basilosaurus*, 'king lizard'. When the vertebrae and associated fragments of the jaw, teeth and other parts of the skeleton were brought to London, the famous anatomist Richard Owen recognized that they belonged to a mammal. He suggested that this animal might have been related to a sirenian, or sea cow.

Lots of fossils of *Basilosaurus* have been found since the nineteenth century, including a complete skeleton at a famous site in Egypt called Wadi Al-Hitan, or Whale Valley. *Basilosaurus* was an 18-metre-long (59 ft) whale (the same length as a bendy bus), not a sea cow, which began swimming in the warm tropical waters of the Eocene Epoch 38 million years ago. It had long, crocodile-like jaws fitted with fearsome teeth. The king lizard was a top predator that fed on bony fish and sharks, and attacked juvenile whales from time to time, according to bite marks on fossil skulls of its victims. Its body was shaped like a giant eel, and *Basilosaurus* swam by undulating its spine up and down, rather than from side to side like a fish. The forward thrust was probably aided by a tailfin, but we cannot be sure about this because the fleshy tissues of the animal were not preserved. Small flippers at the front end worked in steering and balance, but its stunted rear limbs were too weak to play any role in locomotion and may have been used during mating.

The relationship between *Basilosaurus* and artiodactyls such as the hippopotamus and *Indohyus* is apparent from its limbs. Artiodactyls have an ankle bone, called the astragalus, which resembles a pair of pulley wheels stuck together. This operates as a hinge that allows the foot to move up and down, while limiting its side-to-side motion. *Basilosaurus* has this bone in its short hind limbs. The transformation of the descendants of land animals, which looked like *Indohyus*, into *Basilosaurus* and other early whales is akin to the difficulty of comprehending mountain-building when we reckon with the speed of the elevation. The task seems impossible, until we grasp the cumulative creative potential of amassing small changes over millions of years. Thankfully, to guide this exploration we have a superb fossil record, which reveals a catalogue of extinct species that show increasing fitness for the life aquatic.

Pakicetus was a mammal from Pakistan that looked a bit like a wolf, but paddled around in the water, like *Indohyus*, and snapped at fish with its extended jaws. It had close-set eyes and a thickened tail like an otter's, and the anatomy of its skull and inner ear suggests that it was an early whale. *Ambulocetus*, the walking whale, was the size of a sea lion and swam in both fresh and salty water. It had short legs and enlarged hands and feet, with long digits that were probably webbed and functioned as paddles. Related whales that were fully marine were outfitted with ears that worked below the water. With the passage of millions more years, whale eyes moved from the top of the head to the sides, nostrils shifted from the tip to the top of the snout, hind limbs were reduced to stubs, and the seas fairly brimmed with gigantic cetaceans. These prehistoric whales were very small-brained as cetaceans go; whales with the echolocating systems, complex vocalizations and rich social lives that necessitated large brains came later.

The reason for whale evolution is not known, although one might as well question why bats evolved from flightless mammals, birds from non-avian dinosaurs, or amphibians from fish. Behind every transformation lies an unexploited opportunity, a slightly different way of living that affords the species some advantage in gaining food, or mates, or protection, any one of which enhances the likelihood that its stockpile of genes will be bestowed upon future generations. By investing in increasingly aquatic lifestyles, the earliest whales that walked and waded could eat their fill with minimal interference from other mammals. The prospects widened for their descendants who abandoned the security of the land and riverbed and swam away to the sea.

None of this implies that *Pakicetus* was an ancestor of *Ambulocetus*, or that *Basilosaurus* evolved from either of them. Any one of these species, or all three, might have been extinguished by the emergence of a super predator, or by overwhelming competition from other animals, and left no heirs. But each of these fossils shows features of a wider spectrum of organisms from which evolution selected animals that became the ancestors of the modern toothed and baleen whales. *Ambulocetus* was not 'the missing link' between these land and marine mammals, any more than every organism dead or alive served or serves as a link between its ancestors and descendants. The characteristics of species change according to the river of DNA that flows from generation to generation and most forms of life dissolve without a trace as time passes. We are left with a fossil record that allows us to detail a small selection of the animals that arose with modified forms over millions of years – fragments of this Eocene epic in the case of the whales. The task of reconstructing the history of life from fossils is like describing the colours of a rainbow, beginning with red, green and blue, without being able to see the intervening hues as red bleeds into orange and orange into yellow.

Complementing the fossil record, studies of dolphin embryos reveal some of the genetic modifications that must have occurred during the descent of the whales.[5] Like other vertebrate animals, dolphins look like tiny seahorses in their early 'pharyngula' stage of development, with skin folds in the neck and tiny buds in the positions of arms and legs. The development of the limbs is coordinated by a series of genes, including *sonic hedgehog*, named after a character in a video game, which is expressed at the outer edges of the buds. *Sonic hedgehog* is active in the developing forelimbs of dolphins, but silenced in their hindlimbs that wither away. The initiation and contraction of the hindlimbs in dolphins is a clear example of the way that embryological development is founded on a common body plan, which is stretched and compressed, and nipped, tucked and patched to shape everything in the zoological garden.

Early whales such as *Basilosaurus* had tiny legs attached to a pelvis. Legs and pelvis shrank during the evolution of more modern whales, and the pelvis was reduced to a pair of curved bones that serve as anchors for muscles that guide the penis during mating. Rudiments of limb bones are connected to these pelvic bones in some species, too. In very rare cases, dolphins are born with tiny fins in the position of hindlimbs, which shows that at least some of the genetic controls for hindlimb development remain buried in the cetacean genome. Most of the time, as we have seen, these instructions are silenced in the early embryo.

Gene conservation allows evolution to get very playful in some animal groups. Forelimbs were eliminated during the evolution of snakes from lizards, but some species retained hindlimbs of varying sizes for tens of millions of years before they were phased out. By the middle of the Cretaceous Period, 100 million years ago, the back legs of some snakes were only 1 cm or so in length and were attached to a pelvic girdle towards the tail of metre-long bodies. It is conceivable that these teeny limbs helped to propel the animals over obstacles, but it seems more likely that they allowed mates to cling together, rather than intertwining to form a knot as snakes do today. Although all the limbed snakes are extinct, the genes controlling leg development have stayed put in their descendants. This is evident in the early embryos of pythons and boa constrictors that form a pelvic girdle, femurs and rudimentary claws. These structures degenerate before the reptiles hatch, like the limb vestiges of dolphins, but they leave spurs on the surface of the snakes that the males use to stroke and clasp females during courtship and mating. The reduction and adaptation of hindlimbs for mating shows how structures that evolved with one function, namely locomotion on land, can be co-opted for a completely different purpose. The fact that this happened in whales and in snakes is a stunning example of convergence, which is the term for similarities in form and function that come about independently.

Turtles have splashed in the same direction as whales, leaving dry land and freshwater habitats for the sea, at least four separate times across their 260-million-year history. This is surprising when we acknowledge that the whole point of reptilian evolution was the conquest of the land. This was made possible by the evolution of the reptile egg, with its protective membrane, or amnion, that surrounds the embryo. Baby reptiles emerge as muscular hatchlings, looking like miniaturized adults rather than amphibian tadpoles that require extensive remodelling through metamorphosis. When land turtles became sea turtles in the Jurassic Period, they were forced to come ashore to excavate their nests on sandy beaches. Ichthyosaurs avoided the mass slaughter of hatchlings that flap across the sand by giving live birth at sea to youngsters that hatched within their oviducts. Turtles have doubled down on the original egg-laying strategy by limiting oxygen in their oviducts, which halts the development of the embryo until the mother can find the best spot to dig her nest.[6]

Some of the earliest marine turtles were furnished with a hard shell on the underbelly and nothing more than a soft roofing on top. These transitional animals seem cast between unshelled ancestors and completely armoured descendants. Their adult anatomies evoke stages in the developmental process that occurs in the embryos of living turtles, in which the ribs expand and fuse to form flattened plates before hatching. The largest turtle, *Archelon*, lived towards the end of the Cretaceous Period. It weighed close to 2 tonnes, was more than 4 m (13 ft) long, had a leathery shell and hooked beak, and crunched shellfish and crustaceans on the muddy sea floor. This gigantic reptile became extinct before the asteroid impact 66 million years ago, but representatives of all the ancient groups of sea turtles survived this great pruning of the tree of life. Turtles that lived in freshwater habitats made it through the mass extinction, too, suggesting that some combination of the physical hardiness of these reptiles and their slow metabolism kept them going. Since then, all but one of the original turtle groups has disappeared. The sole surviving taxonomic order of 356 species, called the Testudines, encompasses today's sea turtles, snapping turtles, river, pond, mud and wood turtles, terrapins, softshell turtles, land tortoises and side-necked turtles.[7] This assortment of animals illustrates the plasticity of the turtle *Bauplan*, which permitted evolution to refine different versions of these shelled reptiles for every setting.

The ancestors of different species of giant tortoise must have floated between the mainland and volcanic islands, because they are unable to swim. Hatchlings clinging to driftwood may have been the most successful migrants. Giant tortoises were widespread in continental Africa, North and South America, Southern Asia and Indonesia until they were destroyed by human migration and multiplication. Other species thrived on uninhabited islands including Madagascar, Mauritius, the Canary Islands and Malta, but we caught up with them in short order. The only survivors are native to Seychelles and the Galapagos Islands. The intersection between the evolutionary timescale of millions of years and the swift interference of humans highlights our character as the biological equivalent of an asteroid.

As with reptiles, so with plants: the whole point of their invention was the occupation of the land. Land plants evolved about 500 million years ago, at least 100 million years before the first animals with backbones flopped onto the muddy shore like little seals. The plant colonists resembled liverworts and partnered with fungi to prosper on the sludge created by bacteria and algae. The fungi grew inside the cells of these liverworts and spread their filaments into the soil, feeding the plant with dissolved nutrients in return for sugars produced by photosynthesis. (I would be remiss, as a mycologist, if I neglected the opportunity to recast this process as the conquest of the land by fungi, with a supporting role played by plants.) As plants became more accomplished terrestrials, they equipped themselves with internal pipes for circulating water and dissolved sugars up and down the stem. The greening of the land that resulted from these structural innovations was so thorough that photosynthesis flooded the air with oxygen and depleted the carbon dioxide. The lush forests that grew during the Devonian Period drew so much carbon dioxide from the atmosphere that Earth went from greenhouse to icehouse as glaciers spread and sea levels fell.

Moving on, and skipping hundreds of millions of years to the appearance of flowers in the Cretaceous Period, a scattering of plant species worked out how to return to the sea. The rewards for this transition must have been substantial, because it happened at least four times in separate families of plants belonging to the water plantains, or Alismatales. There was something about the anatomy and physiology of these plants that made this possible, since there are no marine roses, buttercups or lilies – only seagrasses, which have no kinship with the grasses on land.

The biggest hindrance to living in the sea is its saltiness, which kills by dehydration. Plants shrivel and die when the sea floods their soils, and saltwater intrusion owing to rising seas is a major threat to crops in coastal areas. Knowledge of the herbicidal effect of salt allowed the hero of John Wyndham's novel *The Day of the Triffids* (1951) to kill the marauding man-eaters by hosing them with seawater in the film adaptation of 1962. Sea turtles cope with their poisonous lifelong baths in the stuff by using specialized glands that empty a saline sludge from the corners of their eyes. Rectal glands do the same job in sharks, and the kidneys of whales work as powerful filters that maintain the same level of salt in their bloodstream as land mammals.[8] Lacking these organs, seagrasses battle dehydration by concentrating protective molecules in their cells that draw as much pure water from the sea as the ocean draws from the plant. Thereby seagrasses maintain an osmotic balance with the surrounding water. The same mechanism is used to maintain hydration by a splendid array of marine seaweeds, which are unrelated to plants, including the fucoid algae of the rocky shore, the giant kelps that form offshore forests, and the gulfweed, *Sargassum*, which floats in the open ocean.

Seawater presents other challenges to plant growth. Land plants inhale and exhale through openings called stomata in their leaves. Stomata do not work underwater, however, and seagrasses have lost them during their evolutionary history; instead, they absorb carbon dioxide dissolved in seawater to carry out photosynthesis. Light is the other resource needed for photosynthesis, and its intensity falls as the depth of the water increases, which explains why seagrasses grow best in shallow water unclouded by sediment. Finally, reproduction in the sea is a problem for plants whose terrestrial ancestors released pollen into the air. The enormous meadows of Neptune grass in the Mediterranean are clones that have spread via horizontal stems or rhizomes rather than seed dispersal. But seagrasses do produce flowers and release strings of elongated pollen grains that bob through the water. These are intercepted by the female parts of the flower, which are coated with a waterproof adhesive, and buoyant fruits are dispersed in the water a few weeks later.[9] The development of salt tolerance and novel methods of reproduction, and the loss of stomatal breathing, was a time-consuming business, but the advantage of a territory unexplored by plants was sufficient stimulus for natural selection to re-equip freshwater species so that they could flourish in the sea. The process was similar to the journey of the whales from land to fresh water, to brackish swamps and, at last, into the open ocean. Fossils of seagrasses show that the job was completed around 100 million years ago.

Because seagrasses look a lot like land grasses to untrained eyes, the sophistication of their adaptations to growing and reproducing in salt water is underappreciated. Whales and sea turtles are more obvious icons of the dynamism of natural selection. Other groups of secondarily marine animals – those with terrestrial ancestors – have equally compelling biographies. In every case, we find parallels in the anatomical overhaul that took place in the development of these species, most obviously in the transformation of limbs made for walking into the flippers of seals, manatees and the extinct groups of marine reptiles.[10] Separated by more than 200 million years, the ancestors of whales and ichthyosaurs left the solid surface of the continents for the fluid of the ocean. Pterosaurs, birds and bats mastered the air through similarly complex modifications of their ancestors, transforming forelimbs into wings rather than flippers.

The working practices of Anthony Trollope, who wrote 47 novels while pursuing a distinguished career in the Post Office, serve as a metaphor for the steadiness of evolution: 'I have been constant, – and constancy in labour will conquer all difficulties.'[11] Natural selection is the author of life. Fin becomes foot and the amphibian pads across the mud; webs are stretched between fingers and toes and the reptile and mammal return to the water; reptile and mammal fingers extend and stretch thin leather wings, and feathers are arrayed along the arms of dinosaurs that become birds; and with a part here and a part there dissolving in the embryo, fresh animals are forged as evolution plays with great gulps of time.

TEN

BEGINNINGS

Billions of Years (10^{16} Seconds)

Biology and geology change apace. Life fashions and refashions itself as Earth's surface is crumpled into mountain ranges and ripped apart along ridges of undersea volcanoes. As the giant jigsaw pieces of crust are rearranged, fossils are cast in stripes of rock pressed from rich muds that pile up on the sea floor and on the bottom of lakes. Biology and geology work in consort, shifting the chemistry and climate of the planet, and the cosmos hits the reset button from time to time by dropping an asteroid on paradise. Most biologists concern themselves with bigger organisms, the muscled and leafy upstarts against the microbial hegemony that has drawn itself out across billions of years. They ignore the amoeba in the room. But life did begin, quite mysteriously, with a single cell, and this is how it will end.

Simplifying the timetable a little, life is 4 billion years old, composite eukaryote cells like ours appeared between 2 and 3 billion years ago, and all biology remained microbiology until 1 billion years ago.[1] Earth formed when a blob of nebula left over from the creation of the Sun coagulated in a propitious orbit, which made the surface not too warm, not too cold, just the right temperature. The young planet was kept molten with meteorites, and the collision with a Mars-sized planet called Theia ejected the material that produced the Moon. As the solar system began to settle down and the bombardment subsided, Earth began to cool, and liquid water pooled on the surface. Life formed soon thereafter. Evidence for this early genesis comes from a range of geological specimens that hold chemical traces of biology, as well as some specks that look a lot like we would expect fossilized cells to look.

The oldest fossils appear in rocks that have been pushed upwards and exposed by the erosion of younger deposits in Canada, Greenland, South Africa and Western Australia. Recognizable cells are found in slabs that have been compressed and heated until they have melted and recrystallized, and chemical traces of life are trapped in mixtures of rocks from different sources that have been fragmented and mashed together before melting. The fossil cells are microscopic filaments preserved in 3.5-billion-year-old chert (a hard, fine-grained rock) in Australia. Purely geological processes can create phantoms that look like cells, but the biological identity of the Australian fossils is confirmed by their enrichment with the lighter isotope of carbon (carbon-12).[2] This chemical imbalance is a definitive signature of microbial activity, which arises when enzymes grab carbon-12 to assemble biomolecules and eschew the heavier isotope, carbon-13.

Even older rocks exposed on the eastern shore of Hudson Bay in Canada contain ruby-red tubes of haematite, or iron oxide, which may have been produced by bacteria.[3] These rocks, called metaconglomerates, were formed from layers of magma spewed from undersea volcanoes and are interspersed with sedimentary deposits speckled with quartz crystals that contain the tubes. Bacteria capable of the wizardry of tube manufacture may have lived around hydrothermal vents and used iron as an energy source. Iron-oxidizing bacteria work in this way today, powering themselves by stripping electrons from ferrous iron, and depositing the waste iron, in its rusty ferric form, on their surface. It seems possible that the haematite tubes formed when this ferric iron crystallized around the ancestors of these microbes. There are some wrinkled threads inside some of the tubes that could be the remnants of the cells that discharged the iron. With age estimates on either side of 4 billion years, the red tubes offer the oldest plausible evidence for life.

Other traces of early life come from chemical imprints that are consistent with a biological origin. Graphite particles embedded in Australian rocks appear to be the same age as the haematite tubes and have a similar carbon-12 signature.[4] The graphite is trapped inside crystals of zircon. Because the graphite sits inside the zircon, it must be the same age as these crystals, and radiometric dating shows that they are 4.1 billion years old. As the analysis of ancient rock samples deepens, we are ditching the persistent tale of a struggle for creation on a sterile planet, and realizing that life seems to have been born as soon as the physical conditions stopped being maximally hideous. Once we entertain this narrative of a prompt debut for life on Earth, the possibility of a universe of extraterrestrials becomes more likely. The only flaw in this judgement is that we have no idea how life happened here. We know that it did – look in the mirror – but not how it began.

The 'How?' question was ignored across the centuries, as long as we believed that gods were creative engines that operated on supernatural principles. The concept of spontaneous generation is implicit in the creation myths of all religious traditions. According to the Qur'an, for example, Allah 'made every living thing from water' (21.30) and 'creates what He will; God has power over everything' (24.45). A foggy agnostic version of this story appealed to me as a child. My Auntie Emma lived in Horncastle, a market town in Lincolnshire, and washed her hair in rainwater filtered from an oak barrel beneath a gutter by her back door. This barrel teemed with mosquito larvae and water fleas. Staring into the depths, it seemed these animals were part of the water itself, conjured by some unseen hand. Gnats that danced in my garden, too small to see clearly but glinting in sunbeams, seemed to be another magical part of nature, spun from the pure chemistry of the air.

Until the seventeenth century, scholarship on the reproduction of insects reached no further than these sorts of childish fancies. Observation and guesswork guided ideas about spontaneous births without recourse to experimental tests. Francesco Redi, an Italian investigator, was the first to challenge the orthodoxy.[5] He demonstrated that insects hatched from eggs and that meat placed in a vase would not become riddled with maggots if flies were excluded by covering the opening with 'a fine Naples veil'.[6] This was a game-changer in the study of biology, but it did nothing to help explain where life came from in the first place, and the gods continued to fill this gap in our understanding and create the inaugural spark.

Sparks were at the forefront of the early research on abiogenesis – the development of life from inanimate materials – in which biological molecules formed in glass vessels charged with simple chemicals and zapped with electrical discharges to simulate lightning. These were the primordial soup experiments of Stanley Miller and Harold Urey at the University of Chicago in the 1950s.[7] Regardless of their cleverness, these Frankensteinian experiments may have hampered progress towards an answer to the question

'How?' The problem with the electrification of soup is that it has been presented in textbooks as a compelling demonstration of the beginning of life, when it is nothing of the sort. Miller and Urey offered some clues about the formation of the building blocks of cells without telling us anything about the assembly of the cell. Too many biology lecturers show a diagram of the famous spark chamber and proceed to evolution without acknowledging the utter obscurity of the voyage from chemistry to biology. This is an example of one of the yawning gaps in science that should be taught as a source of inspiration for young investigators.

Experiments in the tradition of Miller and Urey have continued, and by tweaking the physical conditions and changing the recipe of starter chemicals the resulting list of biological molecules has grown to include most of the amino acids from which proteins are made, along with simple sugars and the coding chemicals in RNA molecules called nucleobases. These findings support the idea that RNA (ribonucleic acid) was formed before DNA (deoxyribonucleic acid) and saw to the formation of proteins – which is the principal function of RNA in cells. RNA also possesses some intriguing properties of self-assembly, so that it might have been capable of copying the information in its own sequence of bases and transmitting this form of genetic information into the future.[8] Other research has shown that some of the chemical reactions that produce key intermediates in metabolism can actually take place without any cells at all. Ferrous iron works as a catalyst in these experiments, taking on the role that enzymes play inside cells, driving a network of reactions in a test tube without any genetic supervision.[9] If this sort of chemistry was going on in the primordial soup, we can imagine its incorporation into a protocell. Metabolic reactions clustered with RNA nudge chemistry closer to biology, so this is a remarkable discovery. The only additional requirement is a membrane to serve as a protective envelope, and we have a cell.

And the cell, defined with a membrane, is the key to life. As the distinguished biochemist Frank Harold said, 'In the beginning was the membrane.'[10] The membrane defines a space in which biochemical reactions can be controlled by gathering the reactants and separating them from the hotchpotch of chemicals floating outside. Once this island of order is itemized through some form of genetic instructions, the vital mote has the potential for reproduction. And when this happens, life can get going in earnest and evolution will assume control. This hurdle is called the Darwinian threshold, and as soon as it is surmounted, the most successful versions of the first microbes will begin to be selected over their weaker cousins. Competition is an inevitable consequence of the capacity for information flow from one generation of cells, or protocells, to the next.

There are similarities between the study of the evolution of the cosmos and that of the origin of cells. Physicists are on firm ground when they discuss the expansion of the universe, but are forced to speculate about the immediate events squashed into the first instant of time. Likewise, biologists have an excellent grasp on the process of evolution, but we are left waving our hands when we attempt to explain the formation of the simplest cell. Physicists have had to develop new mathematical methods to probe the Big Bang, and it seems that a fresh approach may also be necessary to get at the first cells. There might be a very neat answer that everyone has missed. It is possible that someone has figured this out already, by virtue of raw intelligence and lateral thinking, and that they lived and died in obscurity without letting the rest of us know. Or perhaps a scientist working on cell biology or biochemistry is very close to the truth at this moment, but needs to be encouraged to pursue a slightly different tack. This does not seem like the kind of scientific breakthrough that will come as the culmination of experiments on cell metabolism; there is something bigger here. It is a tantalizing prospect.

Another question arises as we pursue this investigation: Why did life develop at all? Or, more explicitly: What made Earth make life, rather than continuing as a purely geological globe? If it was a fluke, an improbable event that materialized from a series of unlikely steps, the likelihood for failure multiplies across time. The probability of creating life in this way is akin to the diminishing odds of throwing consecutive heads in a lengthening series of coin tosses. The only way to make progress is to benefit from a mechanism that favours each small step towards complexity. At first glance, the urge for increasing simplicity, or entropy, seems to count against life, which is built from islands of order. But localized detours from chaos are permissible as long as they are paid for by disorder elsewhere. This is where the Sun serves as our saviour by bathing the flora with more than enough energy for photosynthesis, and mineral water heated by magma allows the brilliance of life to blossom around hydrothermal vents.

Entropy is the governor of time and biology. Arthur Eddington, the British astrophysicist who popularized Einstein's theory of general relativity in the 1920s and 1930s, recognized that time would not exist without entropy. He contrived the metaphor 'time's arrow' to explain this ineluctable relationship: an arrow pointing towards increasing randomness, or entropy, indicates the future; an arrow pointing in the other direction, towards greater order and lower entropy, shows the way towards the past. Entropy is the source of 'this one-way property of time'.[11] Philosophers have developed unconvincing critiques of the arrow, from the entertaining yet ultimately impractical thesis of J.M.E. McTaggart, that time does not exist beyond human imagination, to Henri Bergson's prolix quarrel with Albert Einstein about spacetime.[12] Bergson was concerned with the human experience of time, and saw that this can conflict with the passage of time measured with a clock. This mismatch may be real, but it has no influence on the flight of Eddington's arrow.

The more romantic authors of physics do not help things by arguing for some overarching plasticity of time. If we entertain the possibility of several Big Bangs and Big Crunches, rather than the momentous explosion at the outset of our time, each universe must envelop its own arrow of time. Time may even run backwards once the expansive phase of its universe is over and the cosmos begins to contract. Physicists studying quantum gravity ponder these ideas and encourage the philosophical claim that the familiar passage of time is an illusion. Quantum gravity theories, which are part of the Herculean struggle to unite the physics of Einstein with the world of quantum mechanics, produce a universe in which every location in space experiences its own present and all moments exist at once.[13] Biology is blissfully ignorant of this underlying physical chaos. Life behaves according to a single arrow of time, organizing itself around the challenges of surviving the past, living in the present and working in a manner that makes it likely that genes will flow into the future.

Pursuing the arrow of Eddington, there is no shortage of arguments in favour of the phrase 'Entropy is God', which has many online devotees. As we seek new ideas in the study of the origin of life, it is essential that we do not lose sight of this thermodynamic foundation. Nothing is permissible that contravenes the arrow of time, meaning that it requires a decrease in overall entropy to make it work. It does not matter that the individual cell is swimming upstream, against the arrow, as long as its temporary escape from diffusion is balanced by chaos somewhere else in an interlinked manner. Sunbeams allow sunflowers to bloom and set seed. Gigantic sunfish cruise the ocean feeding on jellyfish and lay 300 million eggs at a go. Selfish apes go about the business of their mornings as their cells burn the fuel distilled from breakfast. There are no unpaid debts in any of these transactions. 'Life is a pure flame and we live by an invisible sun within us,' wrote Thomas Browne in the seventeenth century.[14]

The Nobel Prize-winning biochemist Albert Szent-Györgyi is quoted as saying, 'Life is nothing but an electron looking for a place to rest.'[15] He was referring to the energy that resides in the structure of atoms, which is harvested by oxidation reactions and used to power chemical reactions in cells. The quotation conveys that sense of the inevitability of life. Rather than life figuring out how to develop, it was the innate conditions on the planet that forced life into existence. I tell my students that life offers raw energy a fashionable alternative to the tiresome business of heating rocks and water. These ideas do not take us to an explanation of the cell, but they do provide chemical and physical answers to the question 'Why?', which can be seen as modern evocations of Aristotle's concept of the *pneuma*, or 'vital heat', as the driving force of life. We continue to believe that the first cell emerged from this chemical craftwork *spontaneously*, albeit without divine intervention.

The last whiff of alchemy fades from the laboratory of life when we turn our attention to fossils that are the unambiguous imprints of organisms. The oldest ones look like little worms, which evoke the description of the 'one and the same kind of living filaments' that Erasmus Darwin supposed to have been 'the cause of all organic life'.[16] It is difficult to figure out how these pioneering cells made a living, but analysis of the carbon isotopes concentrated in these 3.5-billion-year-old squiggles suggests that some were solar-powered, or photosynthetic, and others mineral-powered, extracting energy from inorganic compounds. It seems likely that a third group consumed their peers. Microbes continue to live and die in mixed communities today. It is not known where life originated, whether, for example, cells arose in the porous chimneys above hydrothermal vents in the deep ocean, or on the land around mud pots and other kinds of hot spring. Interest in such seemingly unappetizing habitats comes from the abundance of chemical energy in these geothermal hotspots, and the presence of microbes that continue to use the streamlined biochemistry from which more complex forms of life might have evolved.

From these relatively simple origins, the more complex eukaryote types of cell – those furnished with a nucleus containing the chromosomes – evolved from the physical merger of microorganisms with complementary lifestyles. The microbial origin of these cells is apparent from their mitochondria, which are quite clearly modified bacteria. Fossils of seaweeds provide the oldest physical evidence of multicellular eukaryotes, with 1.6-billion-year-old red algae from India and 1-billion-year-old relatives of brown algae from Siberia. Besides these algae, we find billion-year-old fungi in the form of fossilized spores and connecting filaments from the Canadian Arctic.[17] These have a reliable date stamp from radioactive elements trapped in the surrounding sediments. Because fungi and animals share a common ancestor, and we have not found any animal fossils of comparable age, the Arctic spores and filaments are the earliest representatives of the super-grouping of organisms from which mushrooms, yeast and man evolved. These fungi are thought to have lived in an estuary, but we do not know what they fed on. The first fossilized animals, which look like ribbed jellyfish, are 600 million years old, and the simplest land plants evolved 500 million years ago, marking the very recent origins of macrobiology.[18]

Returning to the genesis of it all, the memory of the Lincolnshire rain barrel has stuck with me across the decades. With eyes closed, I can picture the mosquito larvae wriggling through the water and wonder whether the building blocks of life might be spewing today, right now, from some uncharted fountain on the sea floor or in the mud bubbling around a hot spring. The fertility of this secret garden would require a re-imaged timescale of creation, collapsing the millions of years we have thought necessary for the transition from chemistry to biology into an instantaneous and continuing manufacturing process, with complex organic molecules spitting from a fountainhead of hot metallic water. This biochemical brew would be undetectable because it would merge with the same materials produced by the living organisms in the surrounding water. Communities of microbes crowding around the concentrated spume of organic compounds would ensure that the *vita nova* would be devoured at its source. Whole cells may have launched themselves from this chemistry 4 billion years ago, for the simple reason that there was nothing waiting in the wings to eat them. Since the appearance of cells, life and its chemical forge have operated as a closed loop, like the ancient Egyptian ouroboros, the cosmic serpent that eats its own tail.

Eyes reopened to the available facts, the agency of a hidden fountain seems almost as fantastic as the biblical story: 'And God said, Let the earth bring forth the living creature after his kind, cattle, and creeping thing, and beast of the earth after his kind: and it was so' (Genesis 1:24). John Milton pursues the story in *Paradise Lost* with an enthusiasm that is absent from the King James Version:

> The Earth obeyed, and, straight
> Opening her fertile womb teemed at a birth
> Innumerous living creatures, perfect forms,
> Limbed and full-grown: out of the ground up rose. (7.453–6)

Animals tear themselves from the ground: 'The tawny lion, pawing to get free / His hinder parts, then springs, as broke from bonds, / . . . the mole / Rising, the crumbled earth above them threw / In hillocks; the swift stag from underground / Bore up his branching head' (7.464–5, 7.467–9). If we permit Milton's bestiary to substitute for organic molecules and single cells, the urgency and ferocity of the verse may suit the actual birth of life.

Our naivety concerning genesis continues to provide the space for dreams and meditation. Metals leach from the rocks along a short stretch of a woodland stream near my home, and the water turns orange with an oily slick of bacteria that power themselves by rusting iron. This splash of colour is one of the oldest expressions of life. Long before any greening of the land, there was this orange in the water. New organisms with new ways of transforming energy have announced themselves across time by redoxing

Ouroboros, the cosmic serpent or dragon that eats its own tail, as depicted in a drawing by Theodoros Pelecanos in the alchemical tract *Synosius* (1478).

elements and adding their own pigments to the palette of colours in nature, painting a living rainbow around Earth. Pink is another early colour that came from bacteria and has been extracted from billion-year-old Mauritanian shale.[19] These bacteria used their pigments to perform photosynthesis. Descendants of other metabolic pioneers live in our bodies, including cells found in our mouths and intestines that combine hydrogen gas with carbon dioxide and release methane.[20]

'And all amid them stood the tree of life,' wrote Milton of Eden, which applies to the invisible galaxy of microbes as well as to my hypothetical fountain.[21] We live within an ancient sculpture, and it threads throughout us, too. Science might be so very close to solving how it all began, and we would not know it. A similar situation confronted naturalists who were thinking about evolution before Darwin. Afterwards Thomas Henry Huxley remarked, 'My reflection, when I first made myself master of the central idea of the "Origin" was, "How extremely stupid not to have thought of that!" I suppose that Columbus' companions said much the same when he made the egg stand on end.'[22] A Darwin for the twenty-first century is at it now, and there may be an Alfred Russel Wallace closing in on the same answers. Ironically, these investigators may solve how life began as we grapple with the climatic end to humanity.

Whether we are an active part in the circus or have moved to the fossil record, life will continue to organize itself according to the speed of its mechanisms. Fleas will jump in a fraction of a second, and new species of insect will evolve over millions of years. Fast movements and slow transformations cannot be done at different speeds; they play according to the needs of life and the physical constraints imposed by the biosphere. Things might happen differently on another Goldilocks planet, on which slower motion is dictated by greater gravitational force, perhaps, or swifter genetic modifications occur in species irradiated by a more luminous star. Wherever it happens, biology has no effect on time, whose arrow flies onwards at a constant speed, the speed of light, across space. Atoms of uranium decay at precisely the same rate here as in the furthest reaches of the universe. Each organism must string itself along this one-way metric, there being no way back.

The flea jumps swiftly, lives for a few months and reproduces. Its little armoured body is made from atoms that originated in exploding stars, and its origins can be traced to the first insects, and back in time all the way to the first cell. Individuals and their species come and go, whole groups of organisms rise and fall, as the planet churns and quakes. The cell has remained the irreducible unit of life for 100 quadrillion seconds, modified into bacteria and crawling amoebas, and multiplied into bristlecone pines and the bodies of whales, and will persist until the Sun runs out of fuel and melts the biosphere, returning Earth to the purity of its geological beginnings.

REFERENCES

Preface

1 John Milton, 'Sonnet VII', in *Milton: Complete Shorter Poems*, 2nd edn, ed. John Carey (Harlow, 2007), p. 153.
2 Although we are unaware of events that take place on a microsecond timescale, some neurological processes operate at this speed. To determine the direction from which a sound is generated we detect the time that elapses between the arrival of sound waves in our left and right ears. These 'interaural time differences' are on the order of tens to hundreds of microseconds. Barn owls use this mechanism to pinpoint rodents rustling through the grass: Dean V. Buonomano, 'The Biology of Time Across Different Timescales', *Nature Chemical Biology*, III/10 (2007), pp. 594–7.
3 M. H. Herzog, T. Kammer and F. Scharnowski, 'Time Slices: What Is the Duration of a Percept?', *PLoS Biology*, XIV/4 (2016), e1002433.
4 R. J. Irwin and S. C. Purdy, 'The Minimum Detectable Duration of Auditory Signals for Normal and Hearing-impaired Listeners', *Journal of the Acoustical Society of America*, LXXI (1982), pp. 967–74; Clara Suied et al., 'Processing Short Auditory Stimuli: The Rapid Audio Sequential Presentation Paradigm (RASP)', in *Basic Aspects of Hearing*, ed. Brian C. J. Moore et al. (New York, 2013), pp. 443–51; Mayuko Tezuka et al., 'Presentation of Various Tactile Sensations Using Micro-needle Electrotactile Display', *PLoS ONE*, XI/2 (2016), e0148410.
5 Scott Neuman, '1 in 4 Americans Think the Sun Goes Around the Earth, Survey Says', www.npr.org, 14 February 2014.

6 William Blake, 'The Marriage of Heaven and Hell', in *William Blake: Selected Poetry*, ed. W. H. Stevenson (London, 1988), p. 73.

7 The power(s) of ten given in the chapter headings refer to the starting point(s) for the range of timescales that we are examining. For example, Chapter Four, on days, weeks and months, starts at one day, or 8.6×10^4 seconds, which is rounded to 10^5 seconds, and moves on to months, beginning with 2.6×10^6 seconds (rounded to 10^6 seconds) for one month. The timestamp for Chapter Four is given as 10^5 to 10^6 seconds, but extends into the next order of magnitude of seconds when we examine biological rhythms that play out over many months. Chapter Eight, concerning millennia, describes the lifespans of organisms that extend from approximately 1,000 years to a few hundred thousand years, or 3.2×10^{10} seconds to 3.2×10^{12} seconds. The timestamp for Chapter Eight is given as 10^{10} seconds. Chapter One encompasses an enormous range of the fastest biological mechanisms that span five orders of magnitude of time, from 10^{-6} to 10^{-1} seconds.

8 J. McFadden and J. Al-Khalili, *Life on the Edge: The Coming of Age of Quantum Biology* (London, 2014); J. C. Brookes, 'Quantum Effects in Biology: Golden Rule in Enzymes, Olfaction, Photosynthesis and Magnetodetection', *Proceedings of the Royal Society A*, CDLXXIII (2017), 20160822, doi: 10.1098/rspa.2016.0822. The fastest movements described in Chapter One occur in microseconds, or 10^{-6} seconds; quantum processes take place 1 billion times faster, in femtoseconds, or 10^{-15} seconds.

ONE   Ballistics: Fractions of Seconds (10^6–10^1 Seconds)

1 The Portuguese man-of-war, *Physalia physalis*, is a siphonophore rather than a true jellyfish. Siphonophores are integrated colonies of individual animals called zooids; a jellyfish is a single organism.

2 Timm Nüchter et al., 'Nanosecond-scale Kinematics of Nematocyst Discharge', *Current Biology*, XVI/9 (2006), pp. R316–18.

3 The prophecy in Book XI of the *Odyssey*, concerning the death of

Odysseus, is commonly translated as happening 'away from the sea', rather than coming 'from the sea'. The second version is championed by Jonathan S. Burgess in 'The Death of Odysseus in the *Odyssey* and the *Telegony*', *Philologia Antiqua*, VII/7 (2014), pp. 111–22.

4 Gregory S. Gavelis et al., 'Microbial Arms Race: Ballistic "Nematocysts" in Dinoflagellates Represent a New Extreme in Organelle Complexity', *Science Advances*, III (2017), e1602552. These stingers are subcellular weapons, formed in specialized organelles inside the cytoplasm of the dinoflagellate cell. Each jellyfish nematocyst is a whole cell situated on the multicellular tentacles of the animal.

5 J. A. Goodheart and A. E. Bely, 'Sequestration of Nematocysts by Divergent Cnidarian Predators: Mechanism, Function, and Evolution', *Invertebrate Biology*, CXXXVI/1 (2016), pp. 75–91.

6 Yossi Tal et al., 'Continuous Drug Release by Sea Anemone *Nematostella vectensis* Stinging Microcapsules', *Marine Drugs*, XII (2014), pp. 734–45.

7 Nüchter, 'Nanosecond-scale Kinematics'.

8 Nicholas P. Money, *Fungi: A Very Short Introduction* (Oxford, 2016).

9 It is possible that additional power comes from the uncoiling of the connecting tube, which may act like a spring.

10 Because the drummer is using two hands, the woodpecker is actually moving a single set of muscles twice as fast. Tap-dance records are complicated by the difference between speeds averaged over one-minute routines versus short bursts of dancing. Records seem to range between 19 and 38 'tap sounds' per second.

11 I. Siwanowicz and M. Burrows, 'Three Dimensional Reconstruction of Energy Stores for Jumping in Planthoppers and Froghoppers from Confocal Laser Scanning Microscopy', *eLife*, VI (2017), e23824.

12 Water molecules stick together, which is why raindrops form beads on windowpanes. This cohesiveness is overpowered by a drop in hydrostatic pressure in the water dragged by the motion of the shrimp claw, which ruptures the liquid, producing bubbles filled with gas. When the bubbles collapse, they damage the claws of the

shrimp as well as the shells of their prey. Any injury to the shrimp is limited, however, because the animal is furnished with fresh claws each time it moults. For an introduction to the research on the hammer blows struck by mantis shrimp, see Sheila N. Patek, 'The Most Powerful Movements in Biology', *American Scientist*, CIII (2015), pp. 330–37.

13 Coraline Llorens et al., 'The Fern Cavitation Catapult: Mechanism and Design Principles', *Journal of the Royal Society Interface*, XIII (2016), 20150930.

14 Yoël Forterre, 'Slow, Fast, and Furious: Understanding the Physics of Plant Movements', *Journal of Experimental Botany*, LXIV/15 (2013), pp. 4745–60.

15 Rachel Carson, *The Sea Around Us* (New York, 1951), pp. 52–3.

16 Examples include *The Merry Flea Hunt* (1621) by Gerrit van Honthorst, which is part of the collection at the Dayton Art Institute, Ohio, and a beautifully candlelit portrait, *The Flea Catcher* (1638), by Georges de La Tour.

TWO Beats: Seconds (10⁰ Seconds)

1 Benjamin Libet, *Mind Time: The Temporal Factor in Consciousness* (Cambridge, MA, 2005).

2 The first use of 'watch' to refer to a timepiece came in the sixteenth century and derived from earlier applications of the noun to people on watch. In the fifteenth century, *alarum* clocks were used to alert guards to the beginning or end of their watch.

3 It was determined that a 1-metre-long (3 ft) pendulum in a clock at sea level would swing from left to right, and back again, sixty times per minute.

4 The number 60 in base 10 is 10 in base 60. We can count to twelve on one hand by touching each finger with the thumb until we complete three sequences. Each time we reach twelve, we can lift one finger on the other hand until we reach sixty. The convenience

of this counting method is one plausible explanation for the adoption of base 60 by the Sumerians in the third century BC.

5 The heart of a small mammal that beats 400 times per minute beats 400 million times in two years. Blue whales, whose heart beats eight to ten times per minute, are sustained by 400 million heartbeats over eighty years. A human heart that beats seventy times per minute for 81 years contracts and relaxes 3 billion times. The average number of heartbeats for a mammal is closer to 1 billion beats per lifetime. Herbert J. Levine, 'Rest Heart Rate and Life Expectancy', *Journal of the American College of Cardiology*, XXX (1997), pp. 1104–6.

6 Anita Guerrini, 'The Ethics of Animal Experimentation in Seventeenth-century England', *Journal of the History of Ideas*, L (1989), pp. 391–407. The brutality reached a monumental level of horror in the 1660s when Robert Hooke and other gentlemen of London's Royal Society dissected dogs kept vivified with the aid of bellows to inflate their lungs. The eighteenth-century writer Bernard Mandeville dispensed with the Cartesian silliness about the inability of animals to feel pain in *The Fable of the Bees* [1714] (London, 1970), p. 198: 'When a creature has given such convincing and undeniable Proofs of the Terrors upon him, and the Pains and Agonies he feels, is there a follower of *Descartes* so inur'd to Blood, as not to refute, by his Commiseration, the Philosophy of that Vain Reasoner?' Even with modern legislation that protects research animals from the most obvious forms of torture, journals of physiology and animal behaviour are filled with experimental methods that would cause most people to vomit.

7 Flexor muscles seem to play an important role in leg extension in the large (and terrifying) fishing spiders that live in South America, according to T. Weihmann, M. Günther and R. Blickhan, 'Hydraulic Leg Extension Is Not Necessarily the Main Drive in Large Spiders', *Journal of Experimental Biology*, CCXV (2012), pp. 578–93.

8 J. E. Carrel and R. D. Heathcote, 'Heart Rate in Spiders: Influence

of Body Size and Foraging Energetics', *Science*, CXCIII (1976), pp. 148–50.

9 Tin Man, in the Metro-Goldwyn-Mayer film *The Wizard of Oz* (1939), was based on the character Tin Woodman in the book by L. Frank Baum *The Wonderful Wizard of Oz* (Chicago, IL, 1900). Tin Woodman was distraught when he stepped on a beetle (pp. 70–72).

10 There is evidence that some nervous connections may be restored after transplantation. Morcos Awad et al., 'Early Denervation and Later Reinnervation of the Heart Following Cardiac Transplantation: A Review', *Journal of the American Heart Association*, V (2016), e004070.

11 William Wordsworth, 'She Was a Phantom of Delight', in *Poems, in Two Volumes by William Wordsworth, Author of the Lyrical Ballads*, vol. I (London, 1815), pp. 14–15.

12 Dirk Cysarz et al., 'Oscillations of Heart Rate and Respiration Synchronize During Poetry Recitation', *American Journal of Physiology – Heart and Circulation Physiology*, CCLXXXVII (2004), pp. H579–87. The hexameter of the original poems was conserved in the translations.

13 Rodney Merrill, 'English Translations of Homeric Epic in Dactylic Hexameters', *Anabases*, XX (2014), pp. 101–10.

14 L. Sakka, G. Coll and J. Chazal, 'Anatomy and Physiology of Cerebrospinal Fluid', *European Annals of Otorhinolaryngology, Head and Neck Diseases*, CXXVIII (2011), pp. 309–16.

15 Joseph G. Bohlen et al., 'The Male Orgasm: Pelvic Contractions Measured by Anal Probe', *Archives of Sexual Behavior*, IX (1980), pp. 503–21; Joseph G. Bohlen et al., 'The Female Orgasm: Pelvic Contractions', *Archives of Sexual Behavior*, XI (1982), pp. 367–86.

16 William Wordsworth, *The Prelude* [1850] (London, 1995), pp. 60–61.

THREE Bats: Minutes and Hours (10^2–10^3 Seconds)

1 Virgil, *Eclogues, Georgics, Aeneid 1–VI*, Loeb Classical Library 63, trans. Henry Rushton Fairclough, rev. George P. Goold, *Georgics*

vol. III (Cambridge, MA, 1999), pp. 196–7. My choice of translation is quoted more often than the version by Fairclough and Goold, and suits the purpose of this chapter.

2   Thomas Nagel, 'What Is It Like to Be a Bat?', *Philosophical Review*, LXXXIII/4 (1974), pp. 435–50.

3   N. Ulanovsky and C. F. Moss, 'What the Bat's Voice Tells the Bat's Brain', *PNAS*, CV/25 (2008), pp. 8491–8.

4   Nagel, 'What Is It Like to Be a Bat?', p. 438.

5   M. J. Wohlgemuth, J. Luo and C. F. Moss, 'Three-dimensional Auditory Localization in the Echolocating Bat', *Current Opinion in Neurobiology*, XLI (2016), pp. 78–86.

6   L. Thaler and M. A. Goodale, 'Echolocation in Humans: An Overview', *Wiley Interdisciplinary Reviews: Cognitive Science*, VII/6 (2016), pp. 382–93.

7   The sentiment was voiced by Lutz Wiegrebe, and quoted by Emily Underwood, 'How Blind People Use Batlike Sonar', www.sciencemag.org/news, 11 November 2014.

8   Coen P. H. Elemans et al., 'Superfast Muscles Set Maximum Call Rate in Echolocating Bats', *Science*, CCCXXXIII (2011), pp. 1885–8; Lawrence C. Rome et al., 'The Whistle and the Rattle: The Design of Sound Producing Muscles', *PNAS*, XCIII (1996), pp. 8095–100.

9   M. W. Holderied, L. A. Thomas and C. Korine, 'Ultrasound Avoidance by Flying Antlions (Myrmeleontidae)', *Journal of Experimental Biology*, CCXXI (2018), jeb189308.

10 Foraging ants that fall into antlion traps are rescued from time to time by their sisters who respond to a chemical SOS signal and even pull at the antlion's jaws as they close on the victim. This makes sense for a social insect whose workers share most of their genes, so that by aiding a sister, you help to preserve your own genes. There is nothing altruistic about this at all. In line with this logic, injured worker ants are ignored or, perhaps, do not cry for help as they slide helplessly towards the jaws below: Krzysztof Miler, 'Moribund Ants Do Not Call for Help', *PLoS ONE*, XI/3 (2016), e0151925.

11 A. J. Corcoran, J. R. Barber and W. E. Connor, 'Tiger Moth Jams Bat Sonar', *Science*, CCCXXV (2009), pp. 325–7.

12 John Muir, *My First Summer in the Sierra* (Boston, MA, 1911), p. 211.

13 Herman Melville, *Moby-Dick; or, The Whale* [1851] (New York, 1992), pp. 423–4.

14 Kevin Healy et al., 'Metabolic Rate and Body Size Are Linked with Perception of Temporal Information', *Animal Behavior*, LXXXVI (2013), pp. 685–96.

15 See www.medienkunstnetz.de; Douglas Gordon's '24 Hour Psycho' features in Don DeLillo's novel *Point Omega* (New York, 2010).

16 J. Smith-Ferguson and M. Beekman, 'Who Needs a Brain? Slime Moulds, Behavioural Ecology and Minimal Cognition', *Adaptive Behaviour* (2019), doi: 10.1177/1059712319826537.

17 T. Latty and M. Beekman, 'Irrational Decision-making in an Amoeboid Organism: Transitivity and Context-dependent Preferences', *Proceedings of the Royal Society B*, CCLXXVIII (2011), pp. 307–12.

18 Neil A. Bradbury, 'Attention Span During Lectures: 8 Seconds, 10 Minutes, or More?', *Advances in Physiology Education*, XL (2016), pp. 509–13.

FOUR Blossoms: Days, Weeks and Months (10^5–10^6 Seconds)

1 Clock proteins in plants switch one another on and off indirectly by activating and deactivating the genes that code for them: R. G. Foster and L. Kreitzman, *Circadian Rhythms: A Very Short Introduction* (Oxford, 2017), pp. 62–80.

2 Theophrastus, *Enquiry into Plants*, vol. I, trans. Arthur F. Hort, Loeb Classical Library (Cambridge, MA, 1916), pp. 344–5. The admiral was Androsthenes of Thasos, who visited Tylos during his exploration of the Arabian coast.

3 Percy Bysshe Shelley, *Prometheus Unbound: A Lyrical Drama in Four Acts with Other Poems* (London, 1820), pp. 157–73.

4 After the rows of leaflets fold, continued disturbance causes the

plant to allow all four leaves to droop at the end of the stalk, and, lastly, to let the stalk slump when it is really aggravated.

5 The thorn exposure mechanism was explored in a relative of the sensitive plant called the littleleaf sensitive-briar, *Mimosa microphylla*: Thomas Eisner, 'Leaf Folding in a Sensitive Plant: A Defensive Thorn-exposure Mechanism?', *PNAS*, LXXVIII (1981), pp. 402–4. This deterrent cannot work for the water mimosa, *Neptunia oleracea*, which engages in rapid leaflet folding but does not have any thorns. Another species, *Codariocalyx motorius*, known as the dancing plant, waves pairs of little leaflets at the base of its flattened leaves when it is disturbed. The function of this movement is a mystery, but might have something to do with simulating butterflies, in order to discourage real butterflies from depositing their eggs on the plant: Simcha Lev-Yadun, 'The Enigmatic Fast Leaf Rotation in *Desmodium motorium*: Butterfly Mimicry for Defense?', *Plant Signaling and Behavior*, VIII/5 (2013), e24473.

6 Thomas Vaux, *The Poems of Lord Vaux*, ed. Larry P. Vonalt (Denver, CO, 1960), p. 16. Ralph Vaughan Williams set the poem to music with the title 'How Can the Tree But Wither?', and it appears in his *Collected Songs*, vol. II (Oxford, 1993), pp. 33–7. Recordings of this song are a reliable stimulus for this author's tears.

7 Cleve Backster (1924–2013) was a polygraph specialist for the CIA who examined the emotional depths of plants and concluded that they were capable of extrasensory perception.

8 Research into the influence of the microbiome on the gut–brain axis is advancing very swiftly: X. Liang and G. A. FitzGerald, 'Timing the Microbes: The Circadian Rhythm of the Gut Microbiome', *Journal of Biological Rhythms*, XXXII (2017), pp. 505–16; Ana M. Valdes et al., 'Role of the Gut Microbiome in Nutrition and Health', *British Medical Journal*, CCCLXI/ Supplement 1 (2018), pp. 36–44.

9 Daphne Cuvelier et al., 'Rhythms and Community Dynamics of a Hydrothermal Tubeworm Assemblage at Main Endeavour Field

– A Multidisciplinary Deep-sea Observatory Approach', *plos one*, IX/5 (2014), e96924.

10 John Milton, *Paradise Lost*, 2nd edn, ed. Alastair Fowler, Book I, ll. 62–3, pp. 63–4.

11 Daphne Cuvelier et al., 'Biological and Environmental Rhythms in (Dark) Deep-sea Hydrothermal Ecosystems', *Biogeosciences*, XIV (2017), pp. 2955–77.

12 Noga Kronfeld-Schor et al., 'Chronology by Moonlight', *Proceedings of the Royal Society B*, CCLXXX (2013), 20123088; Philip Larkin, *Collected Poems* (London, 2004), p. 144.

13 Aristotle and Pliny the Elder thought the brain was the moistest organ, and that this made our moods subject to the tidal pull of the Moon. Cases of mental disturbance were, therefore, instances of lunacy: E. M. Coles and D. J. Cooke, 'Lunacy: The Relation of the Lunar Phases to Mental Ill-health', *Canadian Journal of Psychiatry*, XXIII/3 (1978), pp. 149–52. In our enlightened age, we find no evidence for any lunar influence on human physiology or behaviour, and, contrary to popular conception, the menstrual cycle has nothing to do with the phases of the Moon.

14 A comparable process of interrupted development takes place in pythons, in which a gene involved in limb development flickers on in the embryo before fizzling out during the first day after the egg is laid.

15 Stanley Finger, 'Descartes and the Pineal Gland in Animals: A Frequent Misinterpretation', *Journal of the History of the Neurosciences*, IV/3–4 (1995), pp. 166–82.

16 A. Damjanovic, S. D. Milovanovic and N. N. Trajanovic, 'Descartes and His Peculiar Sleep Patterns', *Journal of the History of the Neurosciences*, XXIV/4 (2015), pp. 396–407.

17 J. M. Field and M. B. Bonsall, 'The Evolution of Sleep Is Inevitable in a Periodic World', *plos one*, XIII/8 (2018), e0201615. If sleep is necessary for organisms with centralized nervous systems (some sort of brain), one wonders whether intelligent life would evolve on a non-periodic planet bathed continuously with light from a pair of stars.

18 Thomas Nashe, *The Works of Thomas Nashe*, ed. Ronald B.
McKerrow, vol. I (Oxford, 1966), p. 355.

19 Thomas Browne, *Sir Thomas Browne's Works, Including his Life and
Correspondence*, ed. Simon Wilkin, vol. II (London, 1835), p. 112:
'in one dream I can compose a whole comedy, behold the action,
apprehend the jests, and laugh myself awake at the conceits thereof.'
Mr Browne would be entertained by a dream I had some years ago,
of being the best friend of the American actor George Clooney,
who had managed, without my knowledge, to have my eyes injected
with silicone. As we walked through an airport concourse, I noticed
that everyone was laughing at me, some with their hands over their
mouths. George was smiling broadly, nodding at his admirers or
giving them a little wave of the hand. Because I saw through the
implants, I had no idea that my eyes were bulging. Although I had
crafted this screenplay, the reason for the mirth was not revealed
to me until George had me look in a mirror. We were such good
friends that I forgave him immediately and laughed along with
the rest of his entourage as we headed for a weekend in Las Vegas.
Someone, help me.

20 William Butler Yeats, *The Collected Poems of W. B. Yeats*, ed. Richard
J. Finneran (New York, 1989), pp. 203–4. In my book *The Rise of
Yeast: How the Sugar Fungus Shaped Civilization* (Oxford, 2018),
p. 30, I asserted that the fungus *Saccharomyces cerevisiae* was the
primum mobile of civilization. This case rests on the importance
of brewing and baking in the development of sedentism, or human
settlements. Earth's daily rotation is treated as the prime mover
of humanity in a different sense in this book's chapter, as the driver
of the rhythms of our physiology and behaviour.

21 Fredegond Shove, *Poems* (Cambridge, 1956), pp. 21–2. Ralph
Vaughan Williams set the poem to music and it appears in his
Collected Songs, vol. II (Oxford, 1993), pp. 14–20. Recordings of
this piece are a second dependable stimulus for this author's tears;
see n. 6.

FIVE Broods: Years (10⁷ Seconds)

1 Virgil, *Eclogues, Georgics, Aeneid I–VI*, Loeb Classical Library 63, trans. Henry Rushton Fairclough, rev. George P. Goold, *Georgics* vol. III (Cambridge, MA, 1999), pp. 198–9, with my substitution of 'orchards' for 'thickets' in the quotation. Virgil describes cicadas in a similar way in *Eclogues*, ibid., Book II, pp. 32–3.

2 Gaines Kan-Chih Liu, 'Cicadas in Chinese Culture (Including the Silver-fish)', *Osiris*, IX (1950), pp. 275–396.

3 B. W. Sweeney and R. L. Vannote, 'Population Synchrony in Mayflies: A Predator Satiation Hypothesis', *Evolution*, XXXVI/4 (1982), pp. 810–21. The German Renaissance genius Albrecht Dürer (1471–1528) included the insect in an engraving known variously as *The Madonna/Holy Family with the Mayfly/Butterfly/Dragonfly/Locust*. Scholarly claims about the insect being a butterfly, dragonfly or locust are outrageous. Dürer pictured animals and plants with great precision (look at his famous *Stag Beetle* of 1505), and would not have ceded the identity of the mayfly to the imagination of the viewer in this piece. The mayfly – everything about it shouts MAYFLY; look at its 'tail', for starters – has alighted beneath the Virgin Mary, who is holding the infant Jesus and is attended by a very elderly Joseph, who appears to be sleeping off the exhaustion of being an earthly parent to the Messiah; God and the Holy Spirit, pictured as a dove, watch from the clouds above. The artist introduced the mayfly as a symbol of fecundity and resurrection, and we would be remiss if we ignored the obvious comparison made between this etching and a later Dürer woodcut by the scholars Larry Silver and Pamela Smith, in P. H. Smith and P. Findlen, eds, *Merchants and Marvels: Commerce, Science, and Art in Early Modern Europe* (New York, 2002), p. 31: 'the woodcut offers a hieratic theophany through symbolic royal synecdoche, as if in distillation of the actual vision manifested in the *Mayfly* engraving.' Well done if you have any clue what this means.

4 A fungal parasite of periodical cicadas has synchronized its life cycle to the thirteen- and seventeen-year broods without devastating their hosts: J. R. Cooley, D. C. Marshall and K.B.R. Hill, 'A Specialized Fungal Parasite (*Massospora cicadina*) Hijacks the Sexual Signals of Periodical Cicadas (Hemiptera: Cicadae: *Magicicada*)', *Scientific Reports*, VIII (2018), 1432. Males normally mate with female cicadas that respond to their calls by flicking their wings. Males infected by the fungus respond to the calls of other males by flicking their wings like females. Spores of the fungus are transmitted from male to male during the same-sex mating that follows.

5 W. D. Koenig and A. M. Liebhold, 'Avian Predation Pressure as a Potential Driver of Periodical Cicada Cycle Length', *American Naturalist*, CLXXXI (2013), pp. 145–9.

6 Yin Yoshimura, 'The Evolutionary Origins of Periodical Cicadas during Ice Ages', *American Naturalist*, CXLIX (1997), pp. 112–24.

7 Gene Kritsky, *Periodical Cicadas: The Plague and the Puzzle* (Indianapolis, IN, 2004), pp. 10, 16.

8 *Greek Lyric II*, trans. David A. Campbell, Loeb Classical Library (Cambridge, MA, 1988), pp. 204–7.

9 Adolf Portmann, *A Zoologist Looks at Humankind*, trans. Judith Schaefer (New York, 1990).

10 Anna G. Warrener et al., 'A Wider Pelvis Does Not Increase Locomotor Costs in Humans: With Implications for the Evolution of Childbirth', *PLOS ONE*, X/3 (2015), e0118903. In 2018 Sophie Power, a 36-year-old ultramarathoner from London, completed the 166-kilometre (103 mi.) Ultra-Trail du Mont-Blanc in 43 hours and 33 minutes while breastfeeding her three-month-old son at aid stations en route. In 2019 Jasmin Paris, another British runner, was the first woman to win the 431-kilometre (268 mi.) Spine Race along the Pennine Way. She beat the previous record by an astonishing twelve hours and, like Ms Power, expressed breast milk for her baby along the way.

11 Holly M. Dunsworth et al., 'Metabolic Hypothesis for Human Altriciality', *PNAS*, CIX (2012), pp. 15212–16; Caitlin Thurber et al.,

'Extreme Events Reveal an Alimentary Limit on Sustained Maximal Human Energy Expenditure', *Scientific Advances*, V/6 (2019), eaaw0341.

12 Richard A. Thulborn, 'Aestivation among Ornithopod Dinosaurs of the African Tria', *Lethaia*, XI (1978), pp. 185–98.

13 H. C. Fricke, J. Hencecroth and M. E. Hoerner, 'Lowland-upland Migration of Sauropod Dinosaurs during the Late Jurassic', *Nature*, CDLXXX (2011), pp. 513–15.

14 John Milton, 'Sonnet 1', in *Milton: Complete Shorter Poems*, 2nd edn, ed. John Carey (Harlow, 2007), pp. 91–3.

15 Rachel Carson, *The Sense of Wonder* (New York, 1956), pp. 88–9.

16 B. Helm and G. A. Lincoln, 'Circannual Rhythms Anticipate the Earth's Annual Periodicity', in *Biological Timekeeping: Clocks, Rhythms, and Behaviour*, ed. Vinod Kumar (New Delhi, 2017), pp. 545–69.

17 D. M. Anderson and B. A. Keafer, 'An Endogenous Annual Clock in the Toxic Marine Dionflagellate *Gonyaulax tamarensis*', *Nature*, CCCXXV (1987), pp. 616–17.

18   Adrian Bejan, 'Why the Days Seem Shorter as We Get Older', *European Review*, XXVII/2 (2019), pp. 187–94.

SIX Bears: Decades (10^8 Seconds)

1 For my biographers: despite the fact of my three-decade research career as a fungal biologist (first peer-reviewed article on fungi published in 1985, last one in 2016), I have never had any interest in consuming psilocybin. My adolescent fondness for Hawkwind's music did nothing to attract me to the deep narcissism of the psychedelic experience.

2 The song about the astronaut, 'Spirit of the Age', written by Robert Newton Calvert and Dave Brock, appeared on the album *Quark, Strangeness and Charm* (Charisma, 1977). Hawkwind's songs were suitably nihilistic for teenage fans, and conveyed snippets of physics and biology that appealed to proto-scientists like me.

3 Ralph O. Schill, ed., *Water Bears: The Biology of Tardigrades* (Cham, Switzerland, 2018).

4 K. Ingemar Jönsson et al., 'Tardigrades Survive Exposure to Space in Low Orbit', *Current Biology*, XVIII (2009), pp. R729–31.

5 K. I. Jönsson and R. Bertolani, 'Facts and Fiction about Long-term Survival in Tardigrades', *Journal of Zoology*, CCLV (2001), pp. 121–3.

6 C. Ricci and M. Pagani, 'Desiccation of *Panagrolaimus rigidus* (Nematoda): Survival, Reproduction and the Influence on the Internal Clock', *Hydrobiologia*, CCCXLVII (1997), pp. 1–13.

7 R. Margesin and T. Collins, 'Microbial Ecology of the Cryosphere (Glacial and Permafrost Habitats): Current Knowledge', *Applied Microbiology and Biotechnology*, CIII (2019), pp. 2537–49.

8 Anastasia V. Shatilovich et al., 'Viable Nematodes from Late Pleistocene Permafrost of the Kolyma River Lowland', *Doklady Biological Sciences*, CDLXXX (2018), pp. 100–102. The authors did not detail how many culture dishes they set up, nor the number of worms they observed. How long did the worms survive, and did they reproduce? Were eggs visible in the samples at the beginning of the experiment? Were any other organisms recovered from the samples?

9 African lungfish live for about twenty years, whereas the oldest specimens of their Australian cousins are still swimming in their late seventies and may approach a centenary in captivity. Lungfish age is estimated by measuring levels of radioactive carbon (^{14}C) trapped in their scales: Stewart J. Fallon et al., 'Age Structure of the Australian Lungfish (*Neoceratodus forsteri*)', *PLOS ONE*, XIV/1 (2019), e0210168. Australian lungfish do not mummify themselves, so their longevity is based on the same sort of dogged persistence that keeps humans going.

10 Phillipa J. Malpas, 'Dying Is Much More Difficult Than You'd Think: A Death by Dehydration', *Permanente Journal*, XXI (2017), doi:10.7812/TPP/16-148.

11 Ichiro Hori, 'Self-mummified Buddhas in Japan: An Aspect of the

Shugen-dô ("Mountain Asceticism") Sect', *History of Religions*, I/2 (1962), pp. 222–42.

12 Ueda Akinari, *Tales of the Spring Rain Harusame Mongatari*, trans. Barry Jackman (Tokyo, 1975), pp. 73–9.

13 Dominique Singer, 'Human Hibernation for Space Flight: Utopistic Vision or Realistic Possibility?', *Journal of the British Interplanetary Society*, LIX (2006), pp. 139–43; Y. Griko and M. D. Regan, 'Synthetic Torpor: A Method for Safely and Practically Transporting Experimental Animals Aboard Spaceflight Missions to Deep Space', *Life Sciences in Space Research*, XVI (2018), pp. 101–7.

14 Lewis Carroll, *Alice's Adventures in Wonderland AND Through the Looking-glass and What Alice Found There* (London, 1998), p. 61.

15 Petronius, *Satyricon*, trans. Michael Heseltine, rev. Eric H. Warmington, Loeb Classical Library (Cambridge, MA, 1987), pp. 54–5.

16 Kathrin H. Dausmann et al., 'Hibernation in a Tropical Primate', *Nature*, CDXXIX (2004), pp. 825–6.

17 C.-W. Wu and K. B. Storey, 'Life in the Cold: Links between Mammalian Hibernation and Longevity', *Biomolecular Concepts*, VII/1 (2016), pp. 41–52.

18 One of the many problems with placing astronauts in suspended animation is the necessary duration of the deep freeze or drug-induced sleep. Our closest potential Goldilocks planet, Proxima Centauri b, is 4.2 light years away. It would take 78,000 years to get there at the speed of NASA's *New Horizons* probe – 58,000 km/h (36,000 mph) – which visited Pluto in 2015. If we got there and found it dry as a bone, however, we would be forced to climb back into the chiller for another 111,000 years to reach the next one, Barnard's Star b. If we went straight to Bernard's Star b from Earth, we could get there in 111,000 years, because the distance between the two exoplanets is about the same as the distance from Earth to Bernard's Star b.

19 Paul Biegler, 'Sleepyheads: The Surprising Promise of Inducing Torpor', www.cosmosmagazine.com, 6 June 2019.

20 Matteo Cerri et al., 'The Inhibition of Neurons in the Central
    Nervous Pathways for Thermoregulatory Cold Defense Induces
    a Suspended Animation State in the Rat', *Journal of Neuroscience*,
    XXXIII/7 (2013), pp. 2984–93.

21 Samuel Beckett, *Waiting for Godot: A Tragicomedy in Two Acts*
    (New York, 1954), Act II, p. 58.

22 Thomas Browne, *Sir Thomas Browne's Works, Including his Life and
    Correspondence*, ed. Simon Wilkin, vol. III (London, 1835), p. 489.
    On thanatophobia as a foundation of Christianity, Edward Gibbon
    wrote, 'When the promise of eternal happiness was proposed to
    mankind on condition of adopting the faith, and of observing the
    precepts, of the Gospel, it is no wonder that so advantageous an offer
    should have been accepted by great numbers of every religion, of
    every rank, and of every province in the Roman empire'; *The Decline
    and Fall of the Roman Empire*, vol. IV (New York, 1993), p. 513.

23 Aubrey D.N.J. de Grey, 'Escape Velocity: Why the Prospect of
    Extreme Human Life Extension Matters Now', *PLOS Biology*, II/6
    (2004), e187.

24 A clinical study in California suggests that treatment with human
    growth hormone can reverse the natural loss of tissue in the thymus
    gland associated with ageing. The thymus is a critical player in
    the immune system, and the investigators argued that the tissue
    regeneration seen in 51- to 65-year-old patients in this study was
    equivalent to a two-year reversal in ageing. The relationship
    between the vitality of the thymus and overall ageing and longevity
    is unknown, but the study is provocative: Gregory M. Fahy et al.,
    'Reversal of Epigenetic Aging and Immunosenescent Trends in
    Humans', *Aging Cell*, XVIII/6 (2019), e13028.

25 Caleb E. Finch, 'Evolution of the Human Lifespan and Diseases of
    Aging: Roles of Infection, Inflammation, and Nutrition', *PNAS*, CVII
    (2010), pp. 1718–24; Fernando Colchero et al., 'The Emergence of
    Longevous Populations', *PNAS*, CXIII/48 (2016), pp. E7681–90;
    X. Dong, B. Milholland and J. Vig, 'Evidence for a Limit to Human
    Lifespan', *Nature*, DXXXVIII (2016), pp. 257–9.

26 Peter B. Medawar, *An Unsolved Problem in Biology* (London, 1952), p. 13.

27 It is ironic, given the acceptance of pseudoscience, that vaccines, which actually keep us alive, are vilified by an alarming percentage of the population. Asked at a forum in my university if I had any idea what might change the minds of Americans who resist vaccinating their children, I answered, 'How about a smallpox epidemic?', which introduced some levity to the gathering. Update: This retort was weakened in 2020 by the voices of resistance to future vaccination against COVID-19.

28 Browne, *Sir Thomas Browne's Works*, vol. III, p. 491.

29 Pliny, *Natural History*, trans. Harris Rackham, Loeb Classical Library (Cambridge, MA, 1942), Book VII, pp. 618–19. Herman Melville expressed the Epicurean or symmetrical view of mortality in *Mardi and a Voyage Thither* [1849] (New York, 1982), p. 899: 'backward or forward, eternity is the same; already we have been the nothing we dread to be.'

30 Beckett, *Waiting for Godot*, p. 57.

SEVEN Bowheads: Centuries (10⁹ Seconds)

1 Bowheads often sound by arching their backs without displaying their flukes before the descent: G. M. Carroll and J. R. Smithhisler, 'Observations of Bowhead Whales during Spring Migration', *Marine Fisheries Review*, XLII/9 (1980), pp. 80–85.

2 Jeanne Calment's recollections of her introduction to Vincent van Gogh have been cited in many publications and have been blurred by editorial laxity.

3 The lance fragment in the female bowhead came from a model that was patented in 1879: J. C. George and J. R. Bockstoce, 'Two Historical Weapon Fragments as an Aid to Estimating the Longevity and Movements of Bowhead Whales', *Polar Biology*, XXXI/6 (2008), pp. 751–4. This lance was superseded by an improved design in 1885. The possibility that lances of the 1879

model continued to be used by native whalers in the late 1880s serves my purpose in merging the birthdays of the teenage whale and Mlle Calment. A tip from the same model of lance was found in the scapula of a male bowhead killed in the coastal waters of Utqiagvik, Alaska, in 2007.

4 Nikolay Zak, 'Evidence that Jeanne Calment Died in 1934 – Not 1997', *Rejuvenation Research*, XXII/1 (2019), pp. 3–12.

5 John C. George et al., 'Age and Growth Estimates of Bowhead Whales (*Balaena mysticetus*) via Aspartic Acid Racemization', *Canadian Journal of Zoology*, LXXVII (1999), pp. 571–80.

6 Mary Ann Raghanti et al., 'A Comparison of the Cortical Structure of the Bowhead Whale (*Balaena mysticetus*), a Basal Mysticete, with Other Cetaceans', *Anatomical Record*, CCCII (2019), pp. 745–60.

7 It is possible to detect daily growth zones in the otoliths of young fish. Each of these circadian bands is no wider than a bacterial cell in hatchlings. Analysis is complicated further by the deposition of sub-daily increments within the daily growth zones. The patterns of daily and sub-daily rings are affected by the metabolic activity of the fish, which changes according to water temperature, food availability and other environmental variables.

8 The decay of radium-226 leads to an increase in the activity of lead-210 in a sample over the course of a century, until the production of lead-210 is balanced by its decay. This is described as a state of 'secular equilibrium'. The degree of disequilibrium in a sample provides an estimate of its age. In this context, 'secular' refers to a process that occurs slowly over a long period. The adjective derives from the Latin *saeculum*, meaning lifetime.

9 G. E. Fenton, S. A. Short and D. A. Ritz, 'Age Determination of Orange Roughy, *Hoplostethus atlanticus* (Pisces: Trachichthyidae) Using ^{210}Pb:^{226}Ra Disequilibria', *Marine Biology*, CIX (1991), pp. 197–202. A later study verified the long lifespan of roughy by employing a more sensitive radiometric dating method: A. H. Andrews, D. M. Tracey and M. R. Dunn, 'Lead-radium Dating

of Orange Roughy (*Hoplostethus atlanticus*): Validation of a Centenarian Life Span', *Canadian Journal of Fisheries and Aquatic Sciences*, LXVI (2009), pp. 1130–40.

10 Julius Nielsen et al., 'Eye Lens Radiocarbon Reveals Centuries of Longevity in the Greenland Shark (*Somniosus microcephalus*)', *Science*, CCCLIII (2016), pp. 702–4. The maximum estimated age of 392 years was associated with a standard deviation of 120 years.

11 Kara E. Yopak et al., 'Comparative Brain Morphology of the Greenland and Pacific Sleeper Sharks and Its Functional Implications', *Scientific Reports*, IX (2019), article 10,022. The relationship between brain size and behavioural complexity is very complicated. The encephalization of an animal – the ratio of the actual size of the brain to the size predicted for related species of similar body weight – is used as a rough guide to its intelligence. The degree of cortical folding and the number of neurons in particular regions of the brain are other useful metrics.

12 Ugo Zoppi et al., 'Forensic Applications of 14c Bomb-pulse Dating', *Nuclear Instruments and Methods in Physics Research B*, CCXXIII (August 2004), pp. 770–75; Laura Hendricks et al., 'Uncovering Modern Paint Forgeries by Radiocarbon Dating', *PNAS*, CXVI/27 (2019), pp. 13,210–14.

13 T. A. Rafter and G. J. Fergusson, '"Atomic Bomb Effect" – Recent Increase of Carbon-14 Content of the Atmosphere and Biosphere', *Science*, CXXVI (1957), pp. 557–8.

14 Paul G. Butler et al., 'Variability of Marine Climate on the North Icelandic Shelf in a 1,357-year Proxy Archive Based on Growth Increments in the Bivalve *Arctica islandica*', *Palaeogeography, Palaeoclimatology, Palaeoecology*, CCCLXXIII (2013), pp. 141–51; Steve Farrar, 'Ming the Mollusc Holds Secret to Long Life', *Sunday Times* (28 October 2007).

15 Gregor M. Cailliet et al., 'Age Determination and Validation Studies of Marine Fishes: Do Deep-dwellers Live Longer?', *Experimental Gerontology*, XXXVI (2001), pp. 739–64.

16 C. Taylor and A. Pastron, 'Galapagos Tortoises and Sea Turtles

in Gold-rush Era California', *California History*, XCI/2 (2014), pp. 20–39.

17 We do not know what Aesop said, or even if Aesop existed, but the familiar idiom comes from the fable of the tortoise and the hare. The lesson of the tale centres on the rewards of 'sobriety, zeal, and perseverance': *Aesop's Fables*, trans. Laura Gibbs (Oxford, 2002), p. 117.

18 Mice with an average pulse rate of 600 beats per minute use up roughly the same quota of lifelong heartbeats in two or three years. The heart rate of the Greenland shark is at the low end of the pulse range of the giant tortoise, and it lives for twice as long as the reptile. There are differences between these animals and many other species on the order of hundreds of millions of lifetime heartbeats, but there is a definite clustering around the 1 billion mark. This topic is also considered in Chapter Two.

19 Demographers disagree about the existence of a plateau of human mortality, and some claim that the probability of death at the age of 114 is no greater than it is at the age of 105. If this is true, it suggests that there is no upper limit to human longevity: Elisabetta Barbi et al., 'The Plateau of Human Mortality: Demography of Longevity Pioneers', *Science*, CCCLX (2018), pp. 1459–61. Others say that this conclusion rests on faulty analysis of demographic data: Sarah J. Newman, 'Errors as a Primary Cause of Late-life Mortality Deceleration and Plateaus', *PLoS Biology*, XVI/12 (2018), e2006776.

EIGHT Bristlecones: Millennia (10^{10} Seconds)

1 Thomas Mann, *The Magic Mountain*, trans. John E. Woods (New York, 1995), p. 268.

2 Donald R. Curry, 'An Ancient Bristlecone Pine Stand in Eastern Nevada', *Ecology*, XLVI/4 (1965), pp. 564–6. Curry's calamity was an unintended consequence of his desire to demonstrate that a remarkably old stand of pines grew in Nevada, at a time when

more attention was paid to bristlecones in the White Mountains of eastern California.

3 Matthew W. Salzer et al., 'Recent Unprecedented Tree-ring Growth in Bristlecone Pine at the Highest Elevations and Possible Causes', *PNAS*, CVI/48 (2009), pp. 20,348–53.

4 M. W. Salzar, C. L. Pearson and C. H. Baisan, 'Dating the Methuselah Walk Bristlecone Pine Floating Chronologies', *Tree-ring Research*, LXXV/1 (2019), pp. 61–6.

5 R. M. Lanner, 'Why Do Trees Live So Long?', *Ageing Research Reviews*, I/4 (2002), pp. 653–71. There is a lot of guesswork about the mechanisms that support ancient trees, because so much must be learned from inferences about the life of the plant, rather than direct experimentation.

6 Svetlana Yashina et al., 'Regeneration of Whole Fertile Plants from 30,000-y-old Fruit Tissue Buried in Siberian Permafrost', *PNAS*, CIX/10 (2012), pp. 4008–13. Unlike the decay rates of radioactive isotopes, the half-life of DNA and other biological molecules varies according to temperature. The projected half-life of a short span of DNA in bone is 150 years at 25°C (77°F), increasing to 47,000 years at -5°C (23°F): Morten E. Allentoft et al., 'The Half-life of DNA in Bone: Measuring Decay Kinetics in 158 Dated Fossils', *Proceedings of the Royal Society B*, CCLXXIX (2012), pp. 4724–33.

7 M. Grant and J. Mitton, 'Case Study: The Glorious, Golden, and Gigantic Quaking Aspen', *Nature Education Knowledge*, III/10 (2010), p. 40.

8 Sophie Arnaud-Haond et al., 'Implications of Extreme Life Span in Clonal Organisms: Millenary Clones in Meadows of the Threatened Seagrass *Posidonia oceanica*', *PLOS ONE*, VII/2 (2012), e30454; Frank C. Vasek, 'Creosote Bush: Long-lived Clones in the Mojave Desert', *American Journal of Botany*, LXVII/2 (1980), pp. 246–55.

9 Alex Johnson, 'Blackfoot Indian Utilization of the Flora of the Northwestern Great Plains', *Economic Botany*, XXIV/3 (1970), pp. 301–24; William R. Burk, 'Puffball Uses among North

American Indians', *Journal of Ethnobiology*, III/1 (1983),
pp. 55–62.

10 The ancient map lichen in northern Alaska had a diameter of 370
mm. Its age estimate is based on the likely timing of the withdrawal
of the ice sheet that uncovered its boulder. To reach this impressive
diameter in 10,000 years, the lichen would have grown at an average
'speed' of 0.04 mm per year. This is certainly possible. The fastest
map lichens in the Brooks Range grew at a measured 'speed' of
0.35 mm per year over experimental timespans of four to six years,
whereas many living thalli barely grew at all during the same period:
L. A. Haworth, P. E. Calkin and J. M. Ellis, 'Direct Measurement
of Lichen Growth in the Brooks Range, Alaska, U.S.A., and Its
Applications to Lichenometric Dating', *Arctic and Alpine Research*,
XVIII/3 (1986), pp. 289–96; James B. Benedict, 'A Review of
Lichenometric Dating and Its Applications to Archaeology',
American Antiquity, LXXIV/1 (2009), pp. 143–72.

11 Yew trees are likely to be older than the lichens in British
churchyards, although the claim that the Llangernyw yew in Wales
is more than 4,000 years old is disputed by some botanists.

12 Klaus Peter Jochum et al., 'Deep-sea Sponge *Monorhaphis chuni*:
A Potential Palaeoclimate Archive in Ancient Animals', *Chemical
Geology*, CCC–CCCI (2012), pp. 143–51. The spine does not add a
new layer every year, in the manner of growth rings in a tree, but
the water temperature associated with the formation of each layer
can be determined from the ratio of different isotopes of oxygen
(oxygen-18 to oxygen-16), and the ratio of magnesium to calcium.
Both measurements serve as 'palaeothermometers'. As the water
temperature rises, less oxygen-18 is incorporated into the structure
of the spines and the ratio of magnesium to calcium increases.
Seawater temperature has increased in the 20,000 years since the
end of the last ice age, and this is reflected in palaeothermometer
readings taken from the middle of the spine to its surface. This
method suggested that the sponge was 11,000 years old. Subsequent
analysis of the levels of silicon isotopes and the ratio of germanium

to silicon across the radius of the same spine increased the age estimate to 18,000 years: Klaus Peter Jochum et al., 'Whole-ocean Changes in Silica and Ge/Si Ratios during the Last Deglacial Deduced from Long-lived Giant Glass Sponges', *Geophysical Research Letters*, XLIV (2017), pp. 11,555–64.

13 E. Brendan Roark et al., 'Extreme Longevity in Proteinaceous Deep-sea Corals', *PNAS*, CVI/13 (2009), pp. 5204–8.

14 F. Scott Fitzgerald, *Six Tales of the Jazz Age and Other Stories* (New York, 1960), p. 83.

15 Daniel E. Martinez, 'Mortality Patterns Suggest Lack of Senescence in Hydra', *Experimental Gerontology*, XXXIII/3 (1998), pp. 217–55.

NINE Basilosaurs: Millions of Years (10^{13} Seconds)

1 Johannes G. M. Thewissen, *The Walking Whales: From Land to Water in Eight Million Years* (Oakland, CA, 2014).

2 Johannes G. M. Thewissen et al., 'Whales Originated from Aquatic Artiodactyls in the Eocene Epoch of India', *Nature*, CDL (2007), pp. 1190–94.

3 Charles Darwin, *On the Origin of Species by Means of Natural Selection; or, The Preservation of Favoured Races in the Struggle for Life* (London, 1859), p. 282.

4 Ibid., p. 184.

5 Johannes G. M. Thewissen et al., 'Developmental Basis for Hindlimb Loss in Dolphins and Origin of the Cetacean Body Plan', *PNAS*, CIII/22 (2006), pp. 8414–18.

6 Anthony R. Rafferty et al., 'Limited Oxygen Availability in Utero May Constrain the Evolution of Live Birth in Reptiles', *American Naturalist*, CLXXXI/2 (2013), pp. 245–53.

7 This is a greatly simplified list of turtle types. For an immersive and authoritative reading on this subject, see Turtle Taxonomy Working Group, 'Turtles of the World: Annotated Checklist and Atlas of Taxonomy, Synonymy, Distribution, and Conservation Status', 8th edn, *Chelonian Research Monographs,* VII (2017), pp. 1–292.

8 When U.S. President John F. Kennedy spoke about our relationship
to the sea in a speech in September 1962, he evoked the marine
origin of animals: 'I really don't know why it is that all of us are
so committed to the sea, except I think it is because in addition
to the fact that the sea changes and the light changes, and ships
change, it is because we all came from the sea. And it is an
interesting biological fact that all of us have, in our veins the exact
same percentage of salt in our blood that exists in the ocean, and,
therefore, we have salt in our blood, in our sweat, in our tears. We
are tied to the ocean. And when we go back to the sea, whether it
is to sail or to watch it we are going back from whence we came.' If
Kennedy had been correct about the percentage of salt in our blood,
we could drink seawater, but unfortunately the ocean is three times
saltier than blood: 'Water, water, every where, / Nor any drop to
drink,' as Samuel Taylor Coleridge wrote in *The Rime of the Ancient
Mariner* (1834). See www.jfklibrary.org.

9 L. Guerrero-Meseguer, C. Sanz-Lázaro and A. Marín,
'Understanding the Sexual Recruitment of One of the Oldest and
Largest Organisms on Earth, the Seagrass *Posidonia oceanica*', *PLOS
ONE*, XIII/11 (2018), e0207345.

10 N. P. Kelley and N. D. Pyenson, 'Evolutionary Innovation
and Ecology in Marine Tetrapods from the Triassic to the
Anthropocene', *Science*, CCCXLVIII (2015), aaa3717.

11 Anthony Trollope, *An Autobiography* (Oxford, 1999), p. 365.

TEN Beginnings: Billions of Years (10^{16} Seconds)

1 The simplification of the timetable requires small adjustments to
round the numbers to the nearest single significant figure. Earth
is 4.54 ± 0.05 billion years old, which can be rounded to 5 billion
years, and so on.

2 J. William Schopf et al., 'SIMS Analyses of the Oldest Known
Assemblage of Microfossils Document their Taxon-correlated
Carbon Isotope Compositions', *PNAS*, CXV (2018), pp. 53–8. When

bacteria and archaea assimilate carbon atoms from carbon dioxide, they favour the lighter isotope, carbon-12, over the heavier isotope, carbon-13. This happens at the level of the enzymes that catalyse carbon fixation by photosynthesis or chemosynthesis. These enzymes make use of the extra vibrational energy in the bonds between carbon and oxygen in the lighter form of carbon dioxide, which makes them easier to break. Photosynthesis in plants and eukaryotic algae such as giant kelps also results in the accumulation of the lighter isotope. Animal tissues are enriched in carbon-12 because they eat plants or herbivorous animals, and the combustion of fossil fuels is evident from the decrease in the $^{13}C/^{12}C$ ratio (more C-12) in the atmosphere that has been measured since the 1970s: Jonathan R. Dean et al., 'Is There an Isotope Signature of the Anthropocene?', *Anthropocene Review*, I/3 (2014), pp. 276–87.

3 Matthew S. Dodd et al., 'Evidence for Early Life in Earth's Oldest Hydrothermal Vent Precipitates', *Nature*, DXLIII (2017), pp. 60–64.

4 Elizabeth A. Bell et al., 'Potentially Biogenic Carbon Preserved in a 4.1 Billion-year-old Zircon', *PNAS*, CXII/47 (2015), pp. 14,518–21.

5 Emily C. Parke, 'Flies from Meat and Wasps from Trees: Reevaluating Francesco Redi's Spontaneous Generation Experiments', *Studies in History and Philosophy of Biological and Biomedical Sciences*, XLV (2017), pp. 34–42.

6 Francesco Redi, *Experiments on the Generation of Insects*, trans. Mab Bigelow (Chicago, IL, 1909), p. 36.

7 The Miller–Urey experiments were inspired by Aleksandr Oparin (1894–1980), the Soviet biochemist who suggested that life was spawned by a process of chemical evolution in the primordial soup that existed on the early planet.

8 Addy Pross, *What Is Life? How Chemistry Becomes Biology* (Oxford, 2012), pp. 65–8.

9 K. B. Muchowska, S. J. Varma and J. Moran, 'Synthesis and Breakdown of Universal Metabolic Precursors Promoted by Iron', *Nature*, DLXIX (2019), pp. 104–7; K. B. Muchowska, E. Chevallot-Beroux and J. Moran, 'Recreating Ancient Metabolic Pathways

NATURE FAST and NATURE SLOW

before Enzymes', *Bioorganic and Medicinal Chemistry*, XXVII/12 (2019), pp. 2292–7.

10 Franklin Harold is an expert on bioenergetics who played a pivotal role in promoting the work by Peter Mitchell on the synthesis of ATP by the mechanism of chemiosmosis. Mitchell received the Nobel Prize in Chemistry in 1978 for this breakthrough. It was my great fortune to work as a postdoctoral investigator with Harold in Colorado in the 1990s. The memorable phrase was printed on a card and framed above his desk. His book on the origin of cells is filled with wisdom: *In Search of Cell History: The Evolution of Life's Building Blocks* (Chicago, IL, 2014).

11 Arthur S. Eddington, *The Nature of the Physical World* (Cambridge, 1928), p. 69.

12 J.M.E. McTaggart, 'The Unreality of Time', Mind, xvii (1908), pp. 457–73. It is difficult to extract a crystal-clear view of Bergson's critique of relativity and space-time, but Jimena Canales has a good stab at it in her book *The Physicist and the Philosopher: Einstein, Bergson, and the Debate that Changed our Understanding of Time* (Princeton, NJ, 2015). Spacetime is a pivotal concept in physics that has no impact at all on the way life has played out on Earth. Even so, a footnoted sketch seems advisable. Here goes. On a long drive, we are just as likely to refer to the distance of the journey as to the time it will take. We look at the clock and the milometer (odometer) as the road slips beneath the car, and the GPS displays the remaining distance and time based on our speed and the road conditions. Space is ignored when we are sitting in a chair, although we are hurtling through it as Earth spins and orbits the Sun, but we are aware of time, at least periodically, as the clock ticks and the working day drags or flies by. Driving or seated at a desk, space and time are inseparable because we are always *here* as well as *now*, in a particular location in the universe at a particular time. To find ourselves, we need to fix four dimensions: the Cartesian trio, x, y and z for the place, plus t for the when. This makes intuitive sense. Jumping over a stream in the woods, we can feel all four of these

dimensions change if we pay close attention. Position is gauged from our vision and inner-ear gyroscopes, and time can be counted in the seconds that elapse before, during and after the movement. But when we regard time and space from the viewpoint of relativity, things get more complicated, because it is evident that time depends on location and differs from place to place. My now is always going to be slightly different from your now, and if you were to zip off at an immense speed, a gulf would open between my present and yours. You might return gushing with news about the journey and discover that I have been dead for a century. This is the most familiar of the weird ripples of general relativity.

13 Carlo Rovelli, *Reality Is Not What It Seems: The Journey to Quantum Gravity*, trans. S. Carnell and E. Segre (London, 2016), and reviews by Lisa Randall, 'A Physicist's Crash Course in Unpeeling the Universe', *New York Times* (3 March 2017), and Andrew Jaffe, 'The Illusion of Time', *Nature*, DLVI (2018), pp. 304–5.

14 Thomas Browne, *Sir Thomas Browne's Works, Including his Life and Correspondence*, ed. Simon Wilkin, vol. III (London, 1835), p. 494.

15 The source of this bon mot is elusive, but it can be condensed from the following description by Szent-Györgyi: 'Life has learned to catch the electron in the excited state, uncouple it from its partner and let it drop back to the ground state through its biological machinery, utilizing its excess energy for life processes.' This appeared in W. D. McElroy and B. Glass, eds, *A Symposium on Light and Life* (Baltimore, MD, 1961), p. 7.

16 Erasmus Darwin, *Zoonomia; or, The Laws of Organic Life*, vol. I (London, 1794), p. 507.

17 Corentin C. Loron et al., 'Early Fungi from the Proterozoic Era in Arctic Canada', *Nature*, DLXX (2019), pp. 232–5.

18 Animals from the Ediacaran Period, beginning 635 million years ago and ending 542 million years ago, were the first macroscopic organisms. *Dickinsonia* fossils are oval, leaf-like impressions in rocks, with lots of closely packed segments that radiate either side

of a central body axis. The fossils contain fat molecules related to cholesterol, which are distinctive biomarkers for animals: Ilya Bobrovskiy et al., 'Ancient Steroids Establish the Ediacaran Fossil *Dickinsonia* as One of the Earliest Animals', *Science*, CCCLXI (2018), pp. 1246–9. Species of *Dickinsonia* belonged to a diverse fauna in the Ediacaran. These animals are often described as 'a failed evolutionary experiment', which is a rather imperious judgement about a group of organisms that lived for 100 million years. The dismissive verdict is associated with the German paleontologist Adolf Seilacher (1925–2014), who meant it in the sense that there are no living descendants of the Ediacarans.

19 Nur Gueneli et al., '1.1-billion-year-old Porphyrins Establish a Marine Ecosystem Dominated by Bacterial Primary Producers', *PNAS*, CXV/30 (2018), pp. E6978–86.

20 Methane-producing microorganisms, or methanogens, are archaea rather than bacteria. Like bacteria, archaea are prokaryotes, meaning that their cells lack nuclei.

21 I used Milton's verse from *Paradise Lost* (4.218) in an earlier book to describe how biologists have tended to overlook the overriding importance of microorganisms: Nicholas P. Money, *The Amoeba in the Room: Lives of the Microbes* (Oxford, 2014). Microbes are everywhere, but invisible; similarly, the solution to the origin of life may be all around us, but unrecognized.

22 Thomas Henry Huxley, 'On the Reception of the "Origin of Species"', in *The Life and Letters of Charles Darwin, Including an Autobiographical Chapter*, vol. II, ed. F. Darwin (London, 1887), p. 197.

ACKNOWLEDGEMENTS

Some of the themes in the book were supercharged by teaching a seminar at Miami University titled 'The Science and Art of Time', with my colleague Billy Simms. Conversations with my longtime research partner, Mark Fischer, at Mount St Joseph University in Cincinnati, helped me grapple with some issues in cosmology. My engagement with more philosophical questions was enhanced by my participation in the 'Time and Temporality' seminars organized by the John T. Altman Program in the Humanities at Miami in 2019. Far from Ohio, an encouraging conversation with David Moss, MD, founder of the Illumignossi Project in Wisconsin, served as an invaluable stimulus for the single time interval per chapter approach.

These acknowledgements were written during the COVID-19 pandemic, which has caused us to pay more attention to the passage of time. With fewer distractions, one might have anticipated a great lengthening of the day. Instead, the weeks during lockdown seem to fly by, relatively, highlighting the distance between our varying experience of life and the steadiness of time's arrow. Special thanks go to Michael Leaman at Reaktion Books for his unwavering support under these peculiar circumstances.

PHOTO ACKNOWLEDGEMENTS

The author and publishers wish to express their thanks to the below sources of illustrative material and/or permission to reproduce it:

akg-images/Fototeca Gilardi: p. 161; photo Igor Cheri/Shutterstock. com: p. 50; Dennis Kunkel Microscopy/Science Photo Library: p. 133; Eye of Science/Science Photo Library: p. 92; photo courtesy Kathryn M. Fontaine, John R. Cooley and Chris Simon, from 'Evidence for Paternal Leakage in Hybrid Periodical Cicadas (Hemiptera: *Magicicada* spp.)', *PLOS ONE*, 11/9 (2007), e892: p. 80; photo Level12 via Getty Images: p. 20; courtesy Nicholas P. Money: p. 7; photo Gosha Simonov, reproduced with permission: pp. 38–9; photo Nobumichi Tamura/Stocktrek Images via Getty Images: p. 136; Steve Taylor ARPS/Alamy Stock Photo: p. 65; Universal Images Group North America LLC/Alamy Stock Photo: pp. 108–9.

INDEX